The Neural Imagination

Cognitive Approaches to Literature and Culture Series
Edited by Frederick Luis Aldama, Arturo J. Aldama,
and Patrick Colm Hogan

Cognitive Approaches to Literature and Culture includes monographs and
edited volumes that incorporate cutting-edge research in cognitive science, neuroscience, psychology, linguistics, narrative theory, and related
fields, insofar as this research bears on and illuminates cultural phenomena
such as literature, film, drama, music, dance, visual art, digital media, and
comics, among others. The volumes published in this series represent both
specialized scholarship and interdisciplinary investigations that are deeply
sensitive to cultural particularities and grounded in an understanding of
cross-culturally shared emotive and cognitive principles.

The Neural Imagination

AESTHETIC AND NEUROSCIENTIFIC
APPROACHES TO THE ARTS

Irving Massey

UNIVERSITY OF TEXAS PRESS   AUSTIN

Requests for permission to reproduce material from this work should be sent to:
  Permissions
  University of Texas Press
  P.O. Box 7819
  Austin, TX 78713-7819
  www.utexas.edu/utpress/about/bpermission.html

♾ The paper used in this book meets the minimum requirements of ANSI/NISO
Z39.48-1992 (R1997) (Permanence of Paper).

LIBRARY OF CONGRESS CATALOGING-IN-PUBLICATION DATA

Massey, Irving.
   The neural imagination : aesthetic and neuroscientific approaches to the arts /
Irving Massey. — 1st ed.
      p.   cm. — (Cognitive approaches to literature and culture series)
   Includes bibliographical references and index.
  ISBN 978-0-292-72565-2
   1. Neurosciences and the arts.   I. Title.   II. Title: Aesthetic and
neuroscientific approaches to the arts.
   NX180.N48M37   2009
   701'.1—dc22
                                                                    2009013152

*This book is dedicated to
Anne and Joel Huberman*

Contents

Illustrations

Preface

Art and technology have been converging rapidly in the past few years; an important example of this convergence is the alliance of neuroscience with aesthetics, which has produced the new field of neuroaesthetics.

The Neural Imagination examines this alliance. Neuroscience has demonstrated its relevance for aesthetics in several ways. First, it has identified specific neural accompaniments to aesthetic activities of both artists and audiences. One of the means by which such "localizations" are established is through the study of artists (such as Ravel) who have brain lesions. Nowadays, of course, this work is greatly facilitated by the use of brain imaging techniques. The changes in an artist's work after a stroke, for instance, help to localize the constituent elements of artistic production at the neurological level. Thus an injury to a particular area in the right hemisphere might interfere with a musician's appreciation of melodic contour without affecting other aspects of his/her musical abilities, and one might conclude that the injured area has an important role in "processing" melodies. As for audiences, Jonah Lehrer (2007, pp. 141–142) proposes that the 1913 riot at the first performance of Stravinsky's "Rite of Spring" was caused by the effect of unfamiliar sounds on certain neurons which, when overstimulated, precipitate a flood of dopamine that can, in turn, produce symptoms resembling those of schizophrenia. (The example of individuals such as Boris Pavlovich Nikonov, in whom music induced seizures, is perhaps more persuasive [cf. Avanzini, 2003]. I understand that Jock Murray of Halifax, Nova Scotia is studying musicogenic epilepsy.)

A second way in which neuroscience has become relevant for the arts is by providing the content of certain works. So, for instance, the behaviors associated with neurological syndromes such as Tourette's or Asperger's, or the effects of commissurotomy (the division of the two halves of the brain), are now regularly the subject of novels and films. An even more direct contribution of neurology to the arts comes in the form of "readouts" from the brain: the patterns of electrical activity that carry the codes through which we experience our environment are transcribed directly into sound or into visual displays and are then used as the basis of musical compositions or projected into the plastic arts.

On the other hand, the area of motivation—that is, why we want to have aesthetic experiences—has remained relatively opaque to neurological investigation. It is true that some of the events in the brain accompanying the aesthetic response have been localized. Semir Zeki (1999) argues that painters who specialize in certain techniques stimulate particular electrochemical processes that give us satisfaction. Other researchers have identified groups of neurons that seem to respond to beauty as such, or, at least, to whatever the subjects of those experiments consider beautiful. For the most part, though, the aesthetic as a global experience has remained hard to explain, and the motivation for that experience is not easily distinguished from other forms of pleasure. What makes the problem still more difficult, from the point of view of localization, is the fact that some of the most important elements of the aesthetic experience do not have anything to do with beauty or pleasure, but may be primarily strenuous, thought-provoking, or even painful. It then becomes harder to know what we are trying to localize.

Some of the above subjects are mentioned in other works on the psychology of art. What is particularly new in *The Neural Imagination* is the comparison of a neuroscientific approach to the arts with a humanistic approach. If we are to understand the contribution of neuroscience to aesthetics fully, such a comparison is essential. In addition to examining the general problems outlined above, then, this book juxtaposes three essays on the arts that use a neuroscientific approach (Chapters 2, 3, and 4) with a second part in which the differences between a humanistic and a neurological approach are emphasized. Each of the essays in the first group deals with one large area: the visual arts (Chapter 2); literature (Chapter 3); and music (Chapter 4). In these chapters, several hypotheses are advanced.

In the chapter on the visual arts (Chapter 2), I begin with a historical exploration of possible relations between nineteenth-century neuro-psychological theories of form and the rise of cubism. I then go on to modern neurological studies of the visual system in relation to art, ending with the thought that the so-called mirror neurons may have a role in the persistence of representationalism.

In the literature chapter (Chapter 3), I take up the question of dream language and its dissociation from other elements of dreams, the problem of aphasia, the relation between thought and expression, and the bearing of these issues on the language of literature. Ending with the topic of metaphor, I offer the conjecture that the experience of metaphor may be associated with what is called, in psychology, "contrast enhancement" and "the appeal of the rare"—possibly, even, in chemical terms, with an increase in oxytocin levels.

In the music chapter (Chapter 4), I begin with a historical study of the musical dream from Wagner onward. I then go on to explore the implications of this phenomenon for the relations of music, language, and imagery. Finally, I argue that, in song, language is subordinated to musical effects because the language function is relatively weak in the brain's right hemisphere, which tends to be dominant for music.

As I have said, Part 2 of the book, "The Imagination, Plain," emphasizes differences between the humanistic and the neuroscientific approach to the imagination.

For the topic of vision, I work with the observation that forms suggesting closure are generally preferred to open forms, and I compare the impressionistic defense of closed forms by a French nineteenth-century psychologist (G. Calame) with Ramachandran's neurologically informed discussion of this preference, as well as with some recent work by Bar and Meta.

For the area of literature, I explore Patrick Fermor's travelogues as well as what one learns from reading Fermor: namely, that language is not a set of defining rules but a handbook of suggestions. This is not the kind of conclusion that one would reach by using neuroscientific methods alone. Here I emphasize the difference between the kinds of generalizations drawn by literary critics and those drawn by scientists from the study of their materials.

For music, I attempt a new reading of Schubert's song "The Trout," but in this case I conclude that the neurological description of the word-music relation that I offer in Chapter 4 actually supports my impressionistic reading.

Chapter 6, which begins with the "Values" section, uses a variety of references, from Rembrandt to Yiddish poetry, in an attempt to define the tragic moment in art. The purpose of this section is to set up a clear antithesis to any utilitarian conception of art, on which scientific approaches to the arts tend to be predicated. It is intended to establish a clear-cut boundary between what is appropriate for neuroscientific investigation and what answers to a different principle.

In the last sections of Chapter 6, I confront some of the difficult philosophical problems that I encountered during the course of the book and try to reconcile the conflicting positions that emerged during my inquiry. I deal with the question of whether neural events cause or merely accompany mental events and with the uncertain neural status of hypothetical entities (such as "essence," "intentionality," or "imagination" itself).

The final section of Chapter 6, "John Keats to the Rescue," provides a more optimistic perspective. Keats had been a medical student and was aware of the most advanced neurological ideas of his day, but he did not think of neuroscience as a threat to the arts. His poem "Ode to Psyche" is a credo and a manifesto which demonstrates that a reconciliation between neuroscience and the arts is feasible after all. Keats imagines Psyche—once the last but now, by his act, the first among the immortals—as the very goddess of neuroscience. The temple that he will erect for her will not stand among pines, but will be woven of dendrites, "the wreath'd trellis of a working brain," on which the ever-changing buds of imagination will grow. Keats celebrates neuroplasticity: "branched thoughts, new grown with pleasant pain." Scientific thought is welcomed, not feared, by poetry. The only condition for this reconciliation is that it take place within inspiration itself. The primacy of the imagination is the precondition for the appropriate pursuit of neuroscience, of the science of mind; it is only from within the aesthetic that the two can collaborate successfully. Keats' poem, which enshrines Psyche, the soul, as the goddess of neuroscience, is the ideal expression of such a collaboration.

Acknowledgments

I have had the privilege of having my work discussed, as well as criticized, by a number of expert but generous audiences. In 2002 I was invited by Professor Paul Sandor to address the sleep research center at Toronto Western Hospital on the subject of the musical dream. In 2004 Professor Frederick Munschauer had me speak on neuroscience and the arts to the staff of the Jacobs Neurological Institute at Buffalo General Hospital. In 2005 I was kindly invited by Professor Moshe Bar of the Harvard Medical School to talk about my work to his brain-imaging research group. In 2008 Dr. Matthew Woolhouse generously brought me to Cambridge University to speak in his series of seminars on music and science. Various sections of this book have also been presented at conferences or as lectures in Rotterdam, Leeds, Havana, Storrs (Connecticut), Ramat Gan, Boston, and New York.

I wish to thank the following specialists in neurology and its related fields for their willingness to spend time with a layman in the hope, not entirely fulfilled, that he would learn enough to avoid the most egregious errors: Daniel Glaser of the Wellcome Institute; Allan Hobson of Harvard University; Harry Hunt of Brock University; Mark Lythgoe of University College, London; Isabelle Peretz of the Université de Montréal; F. C. Rose of Charing Cross Hospital; Mark Solms of the British Psychoanalytical Society; Jason Warren of University College, London; and Robert Zatorre of McGill University (my alma mater). Matthew Woolhouse of Cambridge University devoted an inordinate amount of time to the scrupulous revision of my essay on the musical dream; the errors that remain are entirely my own. Professor Harry Hunt of the Department of Psychology, Brock University, provided thorough and constructive criticism at several stages of my work. Many other friends and colleagues, especially Professor Richard Abrams and Professor Arthur Efron, assisted me by drawing my attention regularly to publications that I would otherwise have missed. Professors Frederick Aldama, Allan Hobson, Patrick Hogan, Oliver Sacks, and Howard Wolf, as well as Dr. Ephraim Massey, have been unfailingly supportive of my enterprise when my courage waned, as has my editor at the University of Texas Press, Jim Burr. Rachel Massey has provided much thoughtful and discriminating advice on style. I have Professor Ann C. Colley to thank for my title, as well as for those rare moments when common sense prevailed in the preparation of this book. (Her companionship and steady encouragement are, as always, assumed but deeply appreciated.) My trustworthy assistant, Josephine Mariea, has found ways to compensate tactfully for my lack of computer skills. Victoria Davis has done expert work in preparing the index.

The British Library, the Wellcome Library, and Widener Library have all been of great assistance to me in my work. For unfailing reliability, support, and flexible attention over the course of the decades, I am indebted to my mainstay, the Lockwood Library, of the State University of New York at Buffalo, as well as to the music library at the same institution.

It will be apparent that, although this book has required fairly close research, I have not confined my references to articles in peer-reviewed journals. I have felt free to draw my materials from whatever sources I thought helpful without compromising intellectual standards.

Finally, I should like to dedicate this book to Anne and Joel Huberman, both of whom have served *in loco parentis* for my family and myself in emergencies great and small.

Chapter 4, "Music and Language in Dream," is a broadly expanded version of an article titled "The Musical Dream Revisited: Music and Language in Dreams" that appeared in *Psychology of* xvi *Aesthetics, Creativity, and the Arts*, Vol. S1 (2006), 42–50. The relevant passages are reproduced with the kind permission of the American Psychological Association.

The research for Chapters 2, 3, and 4 was supported, in part, by the Department of Neurology, by the Program in Literature and Psychology, by the Julian Park Chair in Comparative Literature, and by the Melodia E. Jones Chair in French, all of the State University of New York at Buffalo. Some of the expenses incidental to the preparation of this book for publication were covered by a generous grant from the Publication Subvention Fund of the College of Arts and Sciences, likewise at the University at Buffalo.

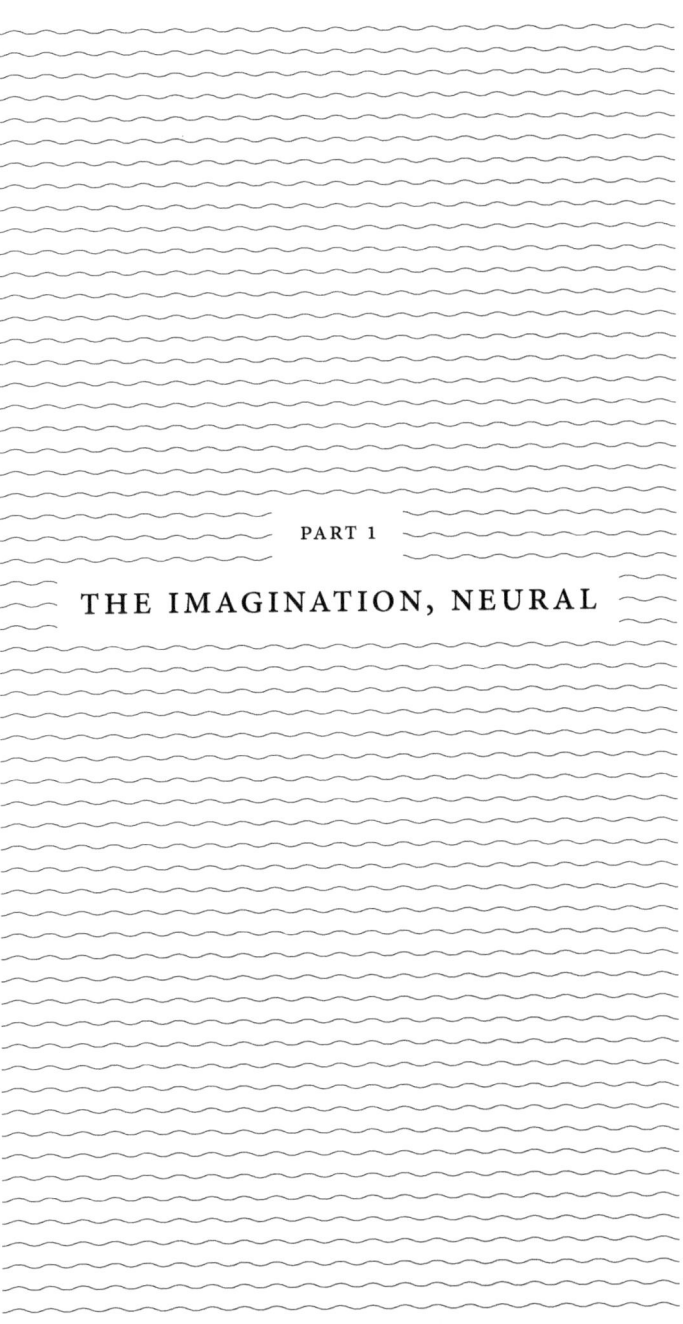

PART 1

THE IMAGINATION, NEURAL

1

Background, Purposes, and Limitations of the Inquiry

1. A BEAUTY SPOT

Much of the neurological investigation into the arts has been concentrated on localization, that is, on the relation of the musical experience of timbre or rhythm, for instance, to a particular set of neurons in a particular region of the brain. There is also the more general question: whether the "beauty response" itself can be localized. If it can, might it then be amenable to electrical inducement? After all, if rats will pedal themselves practically to death by pressing on the "pleasure" bar, and if a subject could be made to find absolutely everything around her intensely funny by having a small area in her medial ventral prefrontal cortex electrically stimulated (Fried et al., 1998; Shammi and Stuss, 1999; and Goel and Dolan, 2001; for the eliciting of a hemi-smile accompanied by euphoria, see Springer et al., 2006), why couldn't one be made to experience beauty automatically, by similar means? Jamie Ward is cited as claiming that beauty is an innate, hardwired response (Garfield, 2006). In fact, Kawabata and Zeki (2004) claim to have located a "beauty spot," a response to beauty in the orbito-frontal cortex, and Vartanian and Goel (2004) have attempted something similar. There is also an article by Cela-Conde et al. (2004), in which the experimenters claim to have found an area in the left dorsolateral prefrontal cortex that shows increased activity when subjects are exposed to pictures that they judge to be beautiful. More recently, Limb and Braun (2008) have done fMRI studies of musicians doing jazz improvisations.

I was much inclined to dismiss all this news as so much cyber-phrenology: finally, what does it tell us about beauty to know that a certain area of our brain shows greater activity when we find something beautiful? After all, in principle, you should be able to localize anything to which you can give a name. At the same time, the case of the artificial inducement of humor, a function almost as subjective and elusive as beauty, left me feeling uneasy.

All this was the last thing on my mind when I went for a walk with my wife last February in the wintry wastes of the so-called Alabama Swamp, about twenty miles outside Buffalo. I was in a bad mood, and I was not enjoying the walk particularly; everything seemed pretty drab. In fact, the swamp is a dreary place, flat and featureless, not exactly what you would think of as a "beauty spot." When, after several hours of walking, we were close to our car again, I started to lag behind. I think, uninteresting as the swamp was, I wasn't eager to get back to the city. There was a small fenced field on my left. Suddenly I noticed a few black teasels high on a bush, then a small tuft of purple-sepia grass reaching out across the snow, and at the back of the little field a mass of yellow reeds like a snowman made out of corn husks. Each of these details was as vivid as if an ophthalmologist had slipped exactly the right lens into the heavy apparatus on the bridge of my nose. I called out to my wife, "I'm being assaulted by inscapes!"

"Inscape" is the word that the poet Gerard Manley Hopkins (1959, p. 127) uses to describe the moment when we catch the essence of what we are perceiving. It is the moment when we do the "quiddity" of our percepts justice. A typical inscape from the Hopkins journals reads,

> there was one light raft of beech which the wind footed and strained on, ruffling the leaves which came out in their trip-lets threaded round with a bright brim like an edge of white ice, the sun sitting at one end of the branch in a pash of soap-sud-coloured gummy bimbeams rowing over the leaves but sometimes flaring out so as to let a blue crust or platter from quite the quick of the orb sail in the eye. (p. 233)

On the drive back to Buffalo, three or four times, in flashing glimpses, I saw other details from the fields on my right with that perfection and intensity, almost arbitrarily, surrounded by meaningless space. It was as though something in me chose to slip the inscape lens into place a few last times before my giving up and going back to the city.

I continue to meditate on the implications of this excursion.

2. PURPOSE OF THE NEUROLOGICAL INVESTIGATION OF THE ARTS

a. Benefits and Limitations

I cannot pretend to shake off entirely the uneasy thoughts aroused by this episode, even as I address the more practical question: What is the purpose of exploring the arts by neurological methods? Perhaps a hint of suspicion haunts the question itself, in the form in which I have phrased it. But what could be more desirable and more innocent than a scientific inquiry, with newly available techniques, into a major area of human activity? The humanists' fear seems to be that the very objectivity of the inquiry, and its proven capacity for correlating the details of psychology with physiological events, will end up subverting the arts, reducing them to ordinary functions without any special privilege. This chapter will begin to explore both the possible advantages and some implied threats of a neurological approach to the arts. I will first mention a few of the obvious benefits in introducing neuroscience to the arts, before allowing these plausible purposes to shade off into objections.

Neuroscience has made tremendous progress in optics in recent years. In consequence, some art schools now find it necessary to offer course modules in neurology, having concluded that it is important to teach students about discoveries concerning vision that may enrich their techniques or otherwise inform their work. There is an annual symposium on art and neuroscience at the Italian Academy in New York. The architectural theorist John Eberhard

(2007) has published a book titled *Architecture and the Brain: A New Knowledge Base from Neuroscience*, which serves as an introduction to neuroscience for architecture students. Such developments may be comparable to the introduction of treatises on perspective during the Renaissance, which may have contributed to the massive changes in representational style that we now take for granted.

Alexander Calder, for instance, was perhaps intuitively aware that adding color to a mobile reduces our awareness of its motion (because we respond more quickly to color than to motion; Zeki, 1999, p. 66), but such an intuitive awareness is quite different from the knowledge of students who have been taught the color/motion principle as a fact and have assimilated it as a standard part of their artistic repertoires. As for music, I have myself, in one of the chapters of this book, attempted a neurological explanation of the ease with which music overrides text in a song setting; whether my explanation, if correct, could contribute to the technique of a composer is another question. On the other hand, David Huron (2006, p. x) claims that from now on composers will indeed have to understand the brain's functions in order themselves to function successfully as composers: "in the absence of knowledge, our only recourse is to follow our intuitions. But intuition is not the foundation for artistic freedom or creative innovation." (One wonders how Chopin or Beethoven managed.)

Let us be clear: Huron is not saying that the composer can use information from cognitive science as a suggestion for possibilities of expression (as chaos theory or fractals have been used in other media, i.e., as metaphors to stimulate the creative mill). Huron apparently thinks that composers need this information in order to understand what they are already doing, so that they may then be able to do it better or differently. Interestingly enough, he ends without offering a single hypothetical example of this imagined process, or what its results might look like.

Data from neuroscience can, of course, be used as *materials* in any medium. Composers such as Horatio Radulescu (2003) and Diego Minciacchi (2003) (not to mention the pop music group

"Heaven 17"), attempt to translate brain data directly into music, by using a special notation. For literature, Lisa Zunshine (2006) shows how dialogue follows the rules of cognitive science in imputing thoughts to an interlocutor, but it is not clear that the formalization of this understanding would help one to write a dialogue. Still, neuroscience has provided an abundance of *subject matter* for any number of novels and films, from *Rain Man* on.

In practical terms, then, some knowledge of neuroscience may be of greater value to the visual artist than to the musician or the author, though the *materials* of neuroscience can furnish subject matter to both, as some artists have used DNA patterns or microbial cultures in their work. (For more on this subject, see Anker and Nelkin, 2004, and Markoff, 2005. It is amusing to see an ad in the April 2006 issue of *Discover* offering DNA variety items for sale.) One of the most promising areas for a beneficial dialogue between neuroscience and the arts is dance. I have been introduced to the work of I. G. Hagendoorn by Daniel Glaser of the Wellcome Institute, and, although dance is a subject about which I do not feel qualified to write, there do seem to be great possibilities in the projects sponsored by Hagendoorn and his group and in Steven Brown's (2008) experiments. But even where no specific use is made of neurological information, no one, and certainly no artist, can help being influenced by the popularization of neuroscientific discoveries and ideas during the last few decades, beginning, perhaps, with Wilder Penfield's stimulating the exposed brain during surgery and producing realistic hallucinations (Penfield and Rasmussen, 1950, p. 36 and *passim*).

There have been numerous observations of what happens to artists after suffering a stroke. One of the most famous cases is that of Ravel, who lost some, but not all, of his musical capacities in consequence of damage to his brain (Mithen, 2005, pp. 53–55; see also Chapter 4, below). Such selective losses of function reveal the existence of what one might call subfunctions; this situation is most familiar in the case of language, where a stroke may deprive one of only the ability to speak, or to understand, or to read, and so forth.

8   The same is true of music, as the area, areas, or networks affected in the brain may serve only one or a few aspects of musical function. What has also been observed is that artists may change their styles after suffering a stroke even if their ability to produce has not been reduced. A well-known example is that of the composer Vissarion Shebalin, again described by Steven Mithen (p. 33). Some physicians specialize in cases of this kind (B. Carey, 2005; Lythgoe et al., 2005; and W. W. Seeley et al., 2008; see also the useful volumes edited by Bogousslavsky and Boller, 2005, and Bogousslavsky and Hennerici, 2007, as well as Chapter 2, Section 2, below).

In view of such observations, it becomes difficult to deny that physiology is destiny; even under the best of circumstances, as Alfred de Vigny ("La Flûte," 1948, p. 201) said, "Des organes mauvais servent l'intelligence" [literally, "Bad organs serve our intelligence"] though it is hard to know what perfect organs would be. For all our attempts to carve out a privileged position for the arts vis-à-vis neurology, when someone acts strangely in a nonaesthetic context, we do not hesitate to assign a physical cause to that person's behavior (see Chapter 6, below). Why should we regard the aesthetic in any other way? If you have a stroke or a brain tumor, it will make you act differently. No one will deny that fact.

Nevertheless, perhaps only as a rear-guard action, I will raise some of the problems associated with such a physiological determinism. For one thing, the expected changes in creative style after someone has suffered brain damage do not always occur (see the article on Visconti and Fellini by Dieguez et al., 2007). (Apparently Balthus' style did not change despite the numerous strokes that the artist suffered.) What is more important, even determined efforts to use the lesion approach to triangulate the source of creative thinking have not yielded satisfactory results. One might assume that the sudden emergence of apparently creative activity after a clearly identifiable injury would enable researchers to locate the origin of the newly expressed abilities. Nevertheless, in their important article, "De novo Artistic Behaviour Following Brain Injury," Pollak et al. (2007) are at pains to dissociate themselves from any confident conclusions concerning the localization of creativity or the combination of neurochemical circumstances that may make creativity possible. As Raymond Tallis (2008, p. 15) remarks, the claim that "the future of aesthetics is in neurology" seems, to say the least, premature.

More broadly speaking, when an artist has a stroke, what changes? Before the stroke, the artist had to work with whatever neurological equipment was at his/her disposal. After the stroke, the artist still has to work with those "organes mauvais," only somewhat more "mauvais" than before. In other words, there is still the assumption of a central "intelligence" or, at least, consciousness, which does what it can with the equipment available to it. I do not see how any number of theories of "distributed processing" or "connectionism" can alter this fact (see Port and Van Gelder, 1995, and Lloyd, 2004).

Again, there is no doubt that neurology can explain certain artistic effects, especially in the plastic arts (see Chapter 2, below). These tend to be of the nature of optical illusions: for instance, Isia Leviant's "Enigma" (Figure 3a), in which concentric circles seem to spin. What neurology cannot do, though, as I will point out again in Chapter 2, is tell us whether "Enigma" is a good or a bad picture, and why. Is it because of its conceptual value, its showing that a physically stationary image is not in fact a fixed image for the brain? Is it because of the particular color scheme that Leviant uses? Is it because of the way the various lines and circles interact? In other words, there are so many elements in and parameters of the picture besides the fact that the circles spin that we are faced in the end with the old, familiar situation in aesthetics, in which a subjective judgment must be arrived at and justified on the basis of a multitude of variables. At that point, neurology has been left behind.

If the purpose of a neurological inquiry is to dispel the "mystery" of the arts in general, then larger objections arise, as I will try to show in Chapter 6. The fact is, there actually are some neuroscientists who would apparently be happy to "explain away" the arts, or who believe that they have already done so (see Kruglinski, 2008;

cf. Ross et al., 2007). I will take up such challenges in more detail below. After my cursory survey of some of the uses, and some of the limitations, of the neurosciences in relation to the arts, I will reconsider a few of the issues that arise in this context in more detail.

b. Mechanical Art: Pros and Cons

In the background of any attempt to bring scientific knowledge to bear on the arts is the fear of ultimate success: that it may be possible to produce a work of art by artificial means. This procedure might consist in having a machine turn out something indistinguishable from the "real thing" or in manipulating the mind through electrochemical techniques so that someone would automatically produce a work of art. If we can make a fruit fly flap its wings by the judicious application of a laser beam, might we not eventually be able to elicit an artistic production by analogous means? To quote Gero Miesenboeck, "If you had a hunch that a certain type of neuron might be important for a particular behavior . . . you could use this technique to see if that's the case—to find out whether activating those neurons elicits the behavior artificially" (quoted in Svoboda, 2006, p. 38). The artist as fruit fly. Or, perhaps, the artist's neurons could be induced to produce art without the artist's participation—or interference. After all, a decapitated frog can perform extraordinarily intricate maneuvers without its head; a male praying mantis, decapitated by its mate, goes on happily with its mating procedures.

At least at first thought, in the aesthetic sphere, these seem to be improbable outcomes. Even if we could determine with what precise parts of the brain Mozart's "Requiem" is associated, could we write another requiem as meaningful as Mozart's by stimulating those areas? Even if an entire work could be generated according to neurological principles, it would have little more than curiosity value. Computer poetry and mixed-media videos (see for instance, Glazier, 2002, and O'Gorman, 2006), computer-generated music, paintings based on optical illusions—all have to have a human impulse behind them in order for us to take them to heart. Even if that impulse should arise from a desire to seek out an impersonal secret in nature (Massey, 1977), it is still only through our conscious mind that that secret will reveal itself. Engaging nature at a more modest level, Jean-Jacques Rousseau ([1781] 1966, p. 58) points out that even for a composer to make us want to hear frogs croaking, the composer must first give them a human voice, must make them sing.

Of course, this whole issue has by now expanded into its own vast field of inquiry and debate. The question of artificial art (a-art) is the subject of books from MIT Press, by Caroline Jones (2006), Mitchell Whitelaw (2004), and others. The Whitelaw book discusses many of the relevant problems in excellent, thoughtful, probing prose, while providing a survey of a-art involving primarily organic forms, materials, and concepts.

One crucial issue it avoids, though, is the necessity of a human investment, in the sense of the artist's committing some important, paid-for element of personal experience to the work. The discussion and the examples are largely formalistic: the book deals mainly with shapes and movements, interactive fields and processes (some, p. 133, involving EEG), and the theories and techniques that produce these results. There is considerable talk about the sociopolitical implications of the a-art movement but very little about the individual artist's experience and what it contributes to the artist's work: sorrow, pain, happiness, and the flavor of an individual's personality and life history, either as playing a role in the product or as recognized by the audience, do not attract much attention in this account.

Another problem, one that Whitelaw does bring up, but only toward the end of the book, might be thought to vitiate the whole a-art enterprise. It is clear that a-art tends toward an ideal horizon at which the machine will produce something that could not by any means have been foreseen by the machine artist. Whitelaw refers to the work of Adrian Thompson (p. 224), a computer scientist whose "breeder" program finally passed beyond his control and began to produce its results in inscrutable ways. If we are to assume that Thompson's results genuinely exemplify "emergence," they might offer the perfect model for an a-art that does not merely reflect, in complicated ways, the program that produced it, for it would have escaped the artist's intentionality entirely (Whitelaw, 2004, p. 226). At the same time, though, it would presumably have lost its special interest for the human observer, becoming a natural phenomenon to which one would respond only as if it were any other object of the senses. (For a sense of the struggle over issues of this sort in information theory, see *Configurations*, 2002.)

The third problem with the panorama of a-art as Whitelaw presents it is the aesthetic as such. The author frequently speaks with approval of particular works, but his comments seem entirely local, suggesting no particular criteria for the appreciation of a-art. The wider question of "What is art?," especially fraught when applied to this body of material, does not receive the treatment that it deserves. It is undoubtedly true that we will be facing an increasingly artificial world. It follows that we will also be facing forms of art increasingly mediated by machines. (It is worth noticing that Ede [2005, pp. 4, 187] still insists that scientists' images, whether diagrams based on DNA or produced according to mathematical formulae, are not *ipso facto* art, yet it is hard to deny an aesthetic dimension to the work of Karen Norberg and her group or to some of the images in Elkins, 2008.) There is a need for an aesthetics that considers the condition of postindustrial civilization as a major feature of our culture as we foresee it.

Besides the general fact that art and the machine are destined to become more closely interrelated, there is also the possibility, as I hint above (in "A Beauty Spot"), that one might eventually be able to induce what one might call a "creative state of mind" by some form of medical-mechanical intervention in neurological processes. After all, something of the sort has been demonstrated with epilepsy, some manifestations of which are similar to aesthetic, spiritual, or religious experiences. Moreover, artists of all kinds have used drugs on the assumption that the creative impulse can be enhanced by artificial means (see Boon, 2002). In the Baudelaire-Rimbaud circle, hashish was favored; of course, it was also understood that the quality of the poetry that was produced depended on who was taking the hashish. By the same token, not everyone as steeped in drugs as Marcel Proust was could be guaranteed to produce an *À la recherche du temps perdu* ([1913–1927] 1966–1969).

The possibility of genetic manipulation must also be considered. (On the genetics of music see Pulli et al., 2008.) An "anxiety gene" in mice has now been isolated, and it is possible to produce courageous "knockout" mice (Asher, 2005), just as, years ago, it was demonstrated that one could produce cowardly bulls by electrical suppression of the amygdala (see "timid bull"). By the way, a specific infection (Toxoplasma gondii) will make rats lose their fear of cats. Might one not be able to isolate a creativity gene? Or at least some neural mechanism that would disinhibit our spontaneous creativity? Perhaps, but it is hard to think that it could be so simple, that all the obstacles to expression could be swept aside by a pill or by a current. The very problems, disasters, and frustrations that block our libido also enter into artistic expression, the complexities and the flavors of which are provided by the fearful visitations of life as much as by its gifts. Assuredly, one could not, without some inner experience of tragedy, have written Hopkins' "Terrible Sonnets," or, for that matter, Shakespeare's *King Lear*. Or Bach's "Komm', süsser Tod."

Yet there are some aspects of creative expression that can have a curiously automatic character about them. I myself have had the experience, though very rarely, of getting into a "tuneful" state at the keyboard, so that I could produce passable tunes, in sequence, without any effort and without any natural limit, as though some melody-producing system had clicked into place for a short time. Had someone observed my brain during that period and stimulated the relevant areas, could the result have been repeated? Although vanishingly rare, those still-better times when music simply flows from the mind or the fingers, not so much as an action but as a state of being, might also be describable at the neurological level, as the lifting of an inhibitory process and the yield of a "productive" gene. In either case, though, the *content* of the expression could not be produced mechanically, since it would have to reflect all the richness and the variety, all the contradictions and the despair, of a particular individual's experience.

c. "Aesthetics": The Term

Before offering further comparisons between humanistic and neuroscientific approaches to the arts, I will attempt to sketch, in its broadest outlines, the field in which this investigation takes place: what is usually called aesthetics. For a general review of the subject, the reader may wish to consult C. Korsmeyer's *Aesthetics: The Big Questions* (1998). The term (which, etymologically, simply means "the study of sensation") began to develop its modern meaning—"the study of beauty or of the arts"—only during the eighteenth century. In that latter sense, it may be divided roughly into three areas: the sources of the art object, the characteristics of the art object, and the relation of the art object to its audience. This distribution (which is admittedly an old-fashioned way of organizing the material) allows in turn for an examination of the relations among these three areas, as in Charles W. Morris' "Esthetics and the Theory of Signs" ([1939] 1971). Most of the issues that arise in critical writing, of whatever school, can be divided in this way, though, of course, the boundaries are often blurred. Even the multiple "Art and" approaches ("Art and Society," "Art and History," "Art and Psychology," "Art and Material Culture," "Art and Medicine," "Art and Linguistics," and so forth, ad infinitum), which bring other disciplines to bear on the arts, can usually be accommodated to this scheme. Strictly speaking, the neuroscientific approach to the arts should not be opposed to the aesthetic one, since, as a member of the "Art and" series, it is as much a part of aesthetics as any other approach.

All of this is manageable within the suggested taxonomy; on the other hand, I think that a satisfactory aesthetics for dealing with the new media, especially with the new mixed electronic media (see "Mechanical Art," above) and bio-art, remains to be developed. Whatever the principles of such an aesthetic turn out to be, I believe that the centrality of motion will be one of them, following the now widespread assumption in psychology that consciousness is largely concerned with phenomena involving motion (cf. Port and Van Gelder, 1995, and Thaut, 2005, p. 25). It gave me quite a shock to walk into the austere immobility of the Chinese sculpture gallery at the Metropolitan Museum of Art after spending a few days with the "Science + Art Festival 2006" in New York City. In her introduction to the anthology *Sensorium: Embodied Experience, Technology, and Contemporary Art*, Caroline A. Jones (2006), takes a first step toward formulating an aesthetics of the contemporary media. For the aesthetics of film, a useful compendium is still Dudley Andrews' *The Major Film Theories: An Introduction* (1976). There are also many books on the theory of conceptual art, among them Charles Harrison's (2001) and Arthur Danto's (2003).

Within the first of the three areas I mention above, that is, the sources of the artwork, we might group such perennial issues as inspiration, the nature of imagination, genius/creativity, and the identity (individual or collective) of the artist. Within the second—the characteristics of the art object—some recurrent topics are: What is art? (E.g., What is the art object in conceptual art, in Tinguely's self-destroying art, in performance art, in Joseph Beuys' sharing a room with a coyote, in mutilating or shooting oneself?) What are the components or structural features of a work—i.e., what are its rhetorical profile and its genre (tragedy, the sublime, etc.)? What are the strategies it deploys? What is tragedy? What is metaphor? What is the aesthetic status of natural objects (e.g., rocks, landscapes)? What are the similarities and differences among the arts? Are all the arts engaged in the same enterprise? What is the relation of form to content in art? What is the role of discord and ugliness in art (e.g., the grotesque)? But here we begin to shade over into the third category, the audience and its relation to the art object. In this area we might place such topics as: For whom does the artist create? (For himself, for others, or for an imaginary, a general, or a specific audience?) Why is there so little agreement about what is good or bad in art? Is taste in art determined by class or social group—i.e., is taste primarily a function of control and power?

There are at least two ways in which neuroscience can be brought to bear on some of the topics treated in discussions of the arts. It can look for neural "sources" for the author's act of production or the audience's act of reception. In other words, it can *localize* these events in the brain. Alternatively, it can take neurological patterns of behavior or response that are known from other situations and demonstrate their *function* in an artistic context. To pick a few random examples: an author who adopts a third-person perspective on a character's behavior activates his/her right inferior parietal cortex (*localization*) (Goldman, 2006, p. 213); the pleasure a metaphor gives us is associated with what psychologists call, in other situations, "the appeal of the rare" (*function*) (see Chapter 3, below); and Monet seascapes shimmer because of the way the artist manipulates the relation between the "what" and the "where" systems, systems that deal with color and luminance (*function*) (see Chapter 2, below).

The problem with these examples is that, in themselves, they have nothing to do with art. This difficulty is fairly obvious with regard to localization: it does not tell us anything about a work of art to say that the work activates, or is associated with, a certain spot in the brain. Functional explanations suffer from a similar defect, though in a less obvious way: they describe the event within the work as if it were just another event in daily life. This strategy is reduced to the ridiculous in such cases as Paul M. Matthews and Jeffrey Mc-Quain's (2003, p. 75) neurological description of Macbeth's reaching for the imaginary dagger as though it were any act of reaching. It remains for traditional aesthetics to grapple with the meaning of the event within the work.

It is actually quite difficult to find areas in which this basic problem does not weaken or undermine neurology's contribution to aesthetics. One such space may be the genetic area, which has to do with the inception of the work of art. It may be possible to imagine setting in motion a process in the brain that would facilitate the production of works of art (see "Mechanical Art," above).

The end product would not be the same as machine-generated art, which always encounters the objection that, in the end, it will be of little interest to live people.

Beyond such possibilities, though, a major value of neuroaesthetics is that it can provide us with a substantially expanded vocabulary for discussing the arts. We can talk in a clearer way about focal vs. peripheral vision and how this opposition affects the way we perceive a painting, even though our understanding of our visual processes may not change our experience of, or our response to, the painting. We can think with a finer discrimination about the varieties of memory instantiated in literature (even if it is only to note the differences among semantic, procedural, and episodic memory) if we know something about theories of memory. We can puzzle in a more hopeful state of mind over the various musical scales and modes that have been developed through the ages if we are aware of the ways in which scales are represented in the brain.

These are no mean achievements. As for the "final" aesthetic questions, those that have to do with the intrinsic nature of art and with the effects of art as art, it may well be argued that traditional aesthetics cannot boast of having had any notable successes in answering these either. The most it can claim is that it has not confused the category of the aesthetic with all the other domains of inquiry that seek to encroach upon it. But we should remind ourselves that, almost by definition, some subjects in traditional aesthetics simply do not lend themselves to investigation by neuroscience or, indeed, by any other system. The wide area of rhetoric and genre studies, for example, is difficult to speak of in terms of other disciplines. More important, it must be remembered that aesthetics is a branch of philosophy, and that there are certain issues within it, such as the definition of art, or the relation of art to truth, that can be addressed only in philosophical terms. Such issues lie largely outside the scope of neuroscience, and there is probably little point in trying to force them into its purview.

What will also be pointed out repeatedly below is that, at least until now, aesthetics has been grounded in acts of immersion in

18 particular works, in an identification with what is happening in those works, and in an appreciation of their distinctive features. (For a brilliant defense of the quiddity of the work of art and the consequent resistance of art to scientific investigation, see Abbott, 2006.) "Humanistic" criticism of the arts tends to drift from observation to observation, rather like a dog inspecting its surroundings, or going from one fire hydrant to the next; it is likely to move from metaphor to metaphor, guided by instinct; it is not required to remain trained on an objective, though it may from time to time come up with an idea, and even a generalization. Again, neuroscience would have no reason to follow such tactics and has little to say about the unique qualities of any single work.

d. The "How" vs. the "Why"

Neurology is, then, of great value in exploring the "how" of aesthetic processes, if not necessarily the "why" or the "what for," or in helping to decide whether one work of art is of greater value than another. Again, neurology is usually more effective in analysis than in synthesis: it can help to localize the components of an aesthetic process, but it is less successful in enabling us to understand how they hang together.

For music, much progress has been made in the localization of rhythm, pitch, melody, and timbre. Some differences have been identified in the ways in which men and women process music. Isabelle Peretz (2003 and 2004) of the University of Montreal has made remarkable discoveries about the amusias (the global or selective failure to respond to music), as well as about the relation of language to music. For the plastic arts, Semir Zeki (1999) and Margaret Livingstone (2002) reveal in extraordinary detail the neurological processes by which we respond to colors, luminance, surface and depth, edges and angles, and straight lines (but not curved lines; Zeki, 1999, p. 116), and they have shown how these features function within actual pictures and other images. For literature, Ramachandran, among others, has attempted to localize the experience of metaphor (Graham, 2005); Lisa Zunshine (2006), as I mention above, examines the ways in which authors show characters imputing motives to other characters, a strategy possibly facilitated by "mirror neurons"; and Ronald Schleifer (2001) links certain aspects of poetry to Tourette's syndrome. For every area of the arts, an attempt has been made to link brain functions to specific features of the art in question. A field of physiological aesthetics, adumbrated in the eighteenth century, has now been firmly established.

Of course, as I have suggested, such inquiries also encounter methodological difficulties and may have inherent limitations. Neuroscience is best at connecting certain features of an aesthetic process with specific events in the brain, but artists and consumers of art care more about the integrated process, the whole experience, than they do about localization. What matters to them is not the science but the phenomenology of the artistic event. To borrow an analogy from the related field of ethics (Tancredi, 2005): even if we localize the neurology of altruism or observe the behavior of some brain area during an ethical dilemma, we still cannot turn over our ethical decisions to a brain function; *somebody* still has to decide what is right. Similarly, as I mention in Chapter 3, a nasal spray containing oxytocin may increase our willingness to trust others (Kosfeld et al., 2005), but we know that there are some people whom we had better not trust. At some point we must be able to take up a position outside our physiology, stand back, and judge for ourselves.

3. SCIENTIFIC VS. HUMANISTIC METHODS

Needless to say, there is not a single "scientific method" or a single "humanistic method." (For one distinguished scientist's summary of scientific research methods, see Changeux, 2004, Chapters 7 and 8.) Both humanists and scientists, of whatever stripe, though, share the same fundamental interest: as Frederick Crews has reminded me, humanists are just as concerned with showing that their opinions are grounded in facts as scientists are. There is a common commitment to reliable information in both pursuits.

Because of this commonality, which should entail some mutual respect, methodological divergences should not lead humanists and scientists to encounter entirely irreconcilable differences when they evaluate the outcomes of their investigations, projects, etc. Still, the methodological differences are significant, and it is my purpose in this chapter to sketch some of them as they relate to this book's project. I will go over this ground again in Chapter 6; the issues are sufficiently important, though, to risk some repetition; they certainly warrant restatement and reformulation. I should also make it clear that, in my subtitle, "Aesthetic and Neuroscientific Approaches to the Arts," I use the word "aesthetic" in its broadest possible acceptance. For my purpose, it comprehends the ideas of the emotional, the subjective, the "plain," the traditional, and the humanistic, as well as such categories as structural analysis and rhetoric. As I mention above, some of these standard topics, such as the relation between art and truth, do not lend themselves to neuroscientific exploration. In "The Imagination, Neural" part of this book, I confine myself to discussing areas, such as the function of metaphor, where there is at least some hope of gaining a foothold in understanding the relation of neuroscience to aesthetic phenomena.

The differences between scientific and humanistic approaches to the arts probably appear obvious to most humanists, and are therefore not in need of reiteration or laborious demonstration. Perhaps this is not equally true for scientists. There has been a steady attempt to encroach on the humanities from the pragmatic, technological, and scientific side ever since Thomas Love Peacock, in "The Four Ages of Poetry," ([1820] 1965) set out two centuries ago to prove that poetry was obsolete. Percy Bysshe Shelley's "Defence of Poetry" ([1821] 1965) was, of course, the classic humanist response. But nowadays, technology threatens to subvert or subsume art at every level, and there are books such as David Edwards' *Artscience:*

Creativity in the Post-Google Generation (2008) which declare unhesitatingly that there is no difference between the two domains. Steven Meyer (2001) argues that, since the beginning of the twentieth century, biological writing has converged with humanistic writing, and that the two have finally become identical. In a similar vein, Ira Livingston (2006, pp. 79–80) remarks, "What was news to biology [in the 1970s] had long been part of the classical legacy of structuralism to those who study language and culture." There is a certain amount of truth to this thesis, but the similarity between aesthetic inquiry and biology can be exaggerated. Of course, both may be concerned with the mind-body problem, but that does not mean that they are the same.

It is partly because of the inescapable differences between science and the humanities that I present Parts 1 and 2 of this book in a somewhat confrontational form. I have no desire to blur the boundary between the two styles of thought; in fact, one of the things I fear is the contamination of humanistic thinking by scientific discourse. H. Porter Abbott (2006, p. 713) remarks that "there are substantial practical and intellectual problems in treating scientific reductionism as a kind of gold standard for all areas of inquiry." J. Adams (2007, p. 204) adds: "The objection to the interest of the sciences in artistic production is founded on a fear that creativity will cede to just such a destructive mechanistic understanding." And, as Tim Dean has astutely pointed out (in conversation), it has even become dangerous to use the term "experimental" when speaking about the arts. Everything has to be "experimental." The term is recognized as honorific; it provides immediate validation, as the prestige of science becomes attached to any attempt at art to which one can relate that term.

In this sense, though certainly not polemic, the object of this book is almost the opposite of Steven Meyer's *Irresistible Dictation: Gertrude Stein and the Correlations of Writing and Science* (2001), as well as of the aforementioned book *Artscience* by David Edwards (2008). Edwards makes a comprehensive case for the fusion of the arts and sciences under the general rubric of creativity, in addition to appealing for the systematic institutional cultivation of inter-disciplinary work between art and science. More insidious is the highly sophisticated suggestion of James Elkins (2008, pp. 99–100) that the border between intuition and calculation cannot be clearly defined (see also Adams, 2007, p. 70).

Actually, the present work is not focused directly on the comparison between the arts and the sciences, since it compares two different discourses about the arts rather than art and science as such. In this context, though, I try to show that, for all the obvious overlap between them, especially in this age of technology, the aesthetic and the scientific approaches to describing the arts represent two different kinds of thinking. I do not wish to reinvent Wilhelm Dilthey's ([1883] 1961) wheel and argue for the irreducible distinction between the humanistic and the natural sciences, or merely to recall the "two solitudes" of C. P. Snow (1959); but I do wish to point out the difficulties in building a genuine bridge between the two areas of inquiry. Such recent commentators as Ede (2005, p. 194) and Barbara Herrnstein Smith (2005, p. 109) believe that these difficulties are insuperable. What is more, they even seem to be happy that they persist. Edelman's (2006) renewed attempt to span the gap between the cultures will not put an end to the debate.

In my previous essays, where I argued, for instance, that metaphor provides access to essence (1977), or that metaphor is an imperfect image seen through a perfect one (1999; see also Chapter 5, Section 3a, below), my purpose was to articulate some generalization derived from a personal experience; there is always some element of autobiography in a humanistic aesthetics. I had no concern whatsoever with the possibility that what I was writing might be expressed in physiological terms or, for that matter, that it might or might not ever be proved (cf. Chapter 6, Section 8, below). Similarly, rather than testability, my ideal in writing at least the second and third sections of Chapter 5 was also accuracy of or fidelity to my own experience, coupled with the capacity to draw out the implications of that experience and to communicate them. Finally, if I could not encapsulate the results of my thinking in a metaphor persuasive of its truth, I would feel that I had fallen short of my purpose. In some sense, the metaphor itself is the proof.

In contrast, when I set out to write the "scientific" essays, I was trying primarily to prove something. Since I am not a practicing scientist, I could not prove anything in my own right, but I could marshal information from accredited scientific sources to support my own hypotheses, such as my supposition that words function differently in song and in speech because song is primarily a right hemisphere function. The intent was to explain an aesthetic observation in physical terms.

To be sure, laboratory experiments may validate hypotheses founded on ordinary experience rather than on technical observations, even in those cases in which scientific confirmation could not have been foreseen (again, see Chapter 6, Section 8, below). In such instances, whereas later proof may be considered a bonus, it is not necessarily the only criterion of value for the original project. Leibniz's "petites perceptions" (1900, vol. 1, p. 80) were shown to exist by Wilder Penfield and Theodore Rasmussen (1950) two centuries later; Theodor Lipps' 1907 hypotheses concerning empathy (as well as Merleau-Ponty's notion of intersubjective awareness [Ostrow, 1990, pp. 42–43]) was confirmed when "mirror neurons" were discovered eighty or so years later; yet neither of these thinkers assumed that his ideas were futile because they lacked laboratory proof. But this situation may be changing.

To get a sense of the gulf that now separates humanistic from "hard" scientific inquiry, one need only read Daniel Dennett's (2005, p. 147) remarks on first-person reporting. Subjective reports about attributes of one's own consciousness (my "personal experiences") are indeed data for the scientist, but the *contents* of those reports are not. (For a dissenting view, see Hurlburt's contribution to Hurlburt and Schwitzgebel, 2007. I am indebted to Allan Hobson for the reference.) To quote Dennett, "*they* [the insights produced by introspection] are not data—the beliefs of subjects about them are the data" (p. 147). Only events that can be verified from a third-person perspective can have the status of scientific data. It would seem that humanists still believe, with Bishop Berkeley, that reality is founded in perception, whereas scientists now believe that it is founded in computation (cf. Lehrer, 2007, p. xii; for an interesting comment, derived from Rothacker and Blumenberg, on the element of subjectivity that cannot be eliminated from signification, see Halmi, 2007, p. 23).

At the level of vocabulary, there are immediate difficulties greater than those that our predecessors, the "natural philosophers," had to confront. A humanistic scholar might have trouble with the following statement: "fusiform and lingual gyri[,] also constitute extrastriate visual association areas" (Jeffries et al., 2003, p. 753). It is hard to know, however, what one could expect a scientist to do with phrases (of mine) such as "perfect vision" or "beauty vs. metaphor," or even with such commonplace judgments as "X was the greatest master of French prose of his time."

I have said that, in setting out to write the first two parts of this book, I began from different positions and with different strategies. These differences in points of departure for the humanistic and the scientific enterprises are usually self-evident; they go far beyond the choice of an inductive or a deductive approach. Obviously, evaluating a text is different from setting up a laboratory experiment, even a "thought experiment." Yet it cannot be denied that there are vast areas of study in the arts, such as "cultural studies," reception theory, or statistical analyses of authorship, that have at least as much claim to scientific status as do the so-called social sciences. These sub-disciplines do not provide the test case, though. For me, a crucial example would be a project that I have in mind as a topic of investigation. I have a sense (I have to say it in that way) that there is something about Patrick Fermor's use of metaphor (see Chapter 5, Section 3b, below) that surpasses our usual understanding of metaphor and reveals something about the "fuzziness" of language itself that we have not yet taken into account. The only way that I can pursue that "hunch" is by immersing myself—or trying to immerse myself—in Fermor's style and sniffing out the special things that are going on in it. Neurology would be of absolutely no use to me in such an effort. Even numerical approaches—e.g., counting words—could serve only to validate or qualify, at a later stage, what intuition alone and submission to the author's influence could ever make apparent. No scientific method could help.

Another, more general example of differences in approach might be taken from the field of memory studies, which are of equal interest to humanists and scientists. The humanist may be observing the broad difference between spontaneous and deliberate recall and its background in Bergsonian philosophy, whereas the scientist may be investigating the contribution of prions to the formation of mnemonic traces. In principle, there may be a way to reduce the first investigation to the second, but a number of obstacles, both technical and systematic, lie in the way. For one thing, the initial observations and hypotheses in scientific psychology have become less and less intuitive or available to the nonspecialist. Even when a first observation has been made at the behavioral level, in terms intelligible to the layman, there is pressure to move as quickly as possible to the physiological level. Here another problem arises: one cannot assume that bridges can be built readily between different orders of function, as each has a certain degree of independence. In "What's Not in your Genes," the biologist H. Allen Orr (2003, p. 40) remarks that some facts are irrelevant: "There are many questions in all fields of science that have little to do with underlying physical detail. . . . [W]e can learn a good deal about the mind . . . without worrying over the details of neurobiology. . . . [T]he answers to such higher-level questions do not change when the underlying mechanisms are discovered." The humanist is not likely to embrace the assumption that, say, human society can be explained by the scientist's molecular processes without the intervention of other explanatory regimes. (For an effort to grapple with the problem of what is often called "emergence" or, in other contexts, "supervenience," see Hogan, 2003, especially pp. 202–210; Modell, 2003, pp. 198–200; Whitelaw, 2004, pp. 207–210; and Adams, 2007, p. 117, 136–137.) Even Elizabeth A. Wilson (2004), who believes that a more flexible neurology could be useful in addressing social questions, acknowledges the difficulty of identifying the intervening stages and structures in such associations. When she asks, rhetorically, "Can historical, cultural, or economic events be serotonergic" (p. 28), the answer has to be sought in the midst of a complex negotiation among the subject, the social conditions, and a neurology that is itself an active player in the historical and affective circumstances it seeks to describe. In fact, Wilson tries to open up the very nature of determination (causation?) to interrogation (p. 26).

One major limitation of humanistic studies is that, although they can deal with mental phenomena and with observable behavior, they have no way of dealing with what goes on in the organism at the sub-behavioral level. This large area of inquiry is closed to humanists. As I have intimated, though, there are also vast areas that either are closed to neuroscientists or simply lie outside the domain of their competence. Take, for example, the large new field of specialization that has developed around mirror neurons, structures within the brain that respond automatically to the expressions on others' faces: if I see fear, the areas activated in my brain are connected with my own fear centers or circuits. However, these discoveries in neuropsychology raise for me major philosophical questions that overflow the domain of science as such. (On the philosophical and social aspects of mimesis, see the interesting book by Huhn, 2004.) Are all our perceptions polluted, so to speak, by an imitative impulse? What would a percept purged of imitation look like? (But even here, some might argue, the neurological study of autism could provide a clue [see Gravitz, 2006]. On the pathological condition known as "imitation behavior," see Tancredi, 2005, p. 70.) Is "perception" just another word for imitation? I find it difficult to think of ways in which a neurologist or even an experimental psychologist could address such a question comprehensively.

As we use language to subdivide experience and our thoughts about experience, we produce entities that do not lend themselves to secure definition but that we still somehow have to deal with. (For the neural status of hypothetical entities, see Chapter 6, Section 6, below.) Whether science can or should follow us into the thickets of onomastic fancy created by our own sorcerer's apprentice is up to scientists to decide. At some point, though, I think the path becomes too difficult, or not worth pursuing.

~~~~~

I write above that the humanist is far more likely than the neuroscientist to study a particular work exhaustively in order to identify its unique characteristics and effect. In Part 2, while comparing the humanistic and neurological approaches to the arts, I emphasize this difference repeatedly. In its most obvious sense, it is simply the difference between the participant's and the observer's point of view, what Smith (2005, p. 110), following anthropological usage, calls the "Etic" and the "Emic," respectively. Although, like the scientist, the humanist brings his/her preconceptions with him/her as a kind of template that isolates and emphasizes certain features of the subject, in the case of the humanist there is always the possibility that the encounter with the individual work, which necessarily overflows any scheme through which it is approached, will tease and draw one out beyond one's preconceptions. The artistic text is a potentially inexhaustible source of insight; one can go on describing its details forever, as if it were a dream. When one immerses oneself in a work of art, the very uniqueness of the work may lead one to, or give rise to, a generalization that only that work could have precipitated, a generalization that, in the absence of that work, would have remained forever unavailable (cf. Chapter 5, Section 3b, below). But the formulation or discovery of a generalization is not the purpose, and certainly not the ulterior purpose, of giving oneself over to the contemplation of a work of art (cf. Chapter 5, Section 3b, below). All one seeks is to do justice to the quiddity of the work; but the mind, in its habit of going beyond the particular, can spin off some broader thought that may have relevance to other situations.

28   To the best of my understanding, the body of fact available to, or produced by, the scientist is not in itself like an art object, which invites contemplation, identification, or appreciation for its own sake and in its own right. The same body of information may yield different theories to different scientists, or to the same scientist at different times, but one does not initially immerse oneself in it for its own sake (cf. Chapter 5, Section 3b, below).

2

Neuroscience and the Visual Arts

"Seeing is believing." Since about a third of our brain is devoted to seeing, it is not surprising that we should like to think that our investment is sound. On the other hand, it is obvious that seeing, especially because it is a mediated sense, taking place at a distance, is uncertain, liable to all sorts of distortions. These distortions (assuming that there is ever a "true" seeing, from which "distorted" seeing deviates) may occur in the light that travels from the object to the eye, in the visual system itself, or in the apparatus of the brain that processes the information that light and the eye provide. Artists play with these possibilities to intensify effects or to diffuse them; in recent times, some have used optical illusions as the basis of their work. Perhaps the most satisfying example of this kind of experimentation is found in the canvases of Monet, who, without resorting to any obvious devices, manages to make viewers feel that some of his "harbor" paintings shimmer and pulse, leaving the eye in a condition of uncertainty as to what it is seeing, or where exactly the things it is seeing are located. Margaret Livingstone (2002) explains this impression neurologically, as a contradiction between the effects of luminance and color that we normally expect.

1. NINETEENTH-CENTURY PSYCHOLOGY, "EMPATHY," AND THE ORIGINS OF CUBISM

The nineteenth century was a period of active psychological study and experimentation in the field of optics, and the aesthetics of vi-

30   sion was a major preoccupation of art criticism during that time. This concern, and particularly a concern with geometrical and aesthetic form, is especially striking in the German psychologists and aestheticians. It is possible to trace the major developments in late-nineteenth-century art—especially the revolt against representation, the emergence of cubism, and, in architecture, the *Bauhaus* movement and *die neue Sachlichkeit* ["the new realism"]—directly to nineteenth-century psychological theories of vision. What I cannot account for is the fact that these ideas, developed mainly in Germany, first came to fruition in French art. I am not alone in my perplexity; Professor Pepe Karmel of New York University tells me that his own research in this field has failed to turn up such an intercultural link. Lynn Gamwell (2002, p. 315) suggests that Cézanne and Gestalt have a common origin in Helmholtz, but she provides no evidence with respect to Cézanne. Charles Edward Gauss, in *The Aesthetic Theories of French Artists 1855 to the Present* (1949, pp. 49–50), tries to account for the development of French cubism in terms of the general intellectual movements of the time, with analogies to Poincaré, Einstein, and new theories in logic and musical composition.

    Of course, there had always been a deep interest in the elementary geometrical forms, from antiquity through the Neo-Platonists and the Renaissance, up to the nineteenth century, when it culminated in the work of Humbert de Superville and Charles Blanc (Stafford, 2007, pp. 32–33). Still, I find it hard to believe that there is no connection, by whatever devious route or intermediaries, between Theodor Lipps' exhaustive study of the elementary shapes in his 1891 essay "Aesthetische Faktoren der Raumanschauung" ["Aesthetic Factors in our Perception of Space"], later expanded to a book titled *Raumaesthetik* [*Aesthetics of Space*], and Cézanne's "manifesto" of 1904, which demands that nature be interpreted through the cylinder, the sphere, the cone, and the plane, almost as a religious obligation (Gauss, 1949, p. 46). (Interestingly enough, enthusiasm for the Platonic shapes in art was strongest among eighteenth-century neoclassicists such as Claude-Nicolas Ledoux and

Étienne Louis Boullée.) Lipps undertook his work in order to show how stresses in perception lead to optical illusions. In fact, Lipps published another work, *Ueber einfachste Formen der Raumkunst* [*Concerning the Simplest Shapes of Spatial Art*] ([1904] 1909) in the same year that Cézanne published his brief manifesto. In *Ueber einfachste Formen der Raumkunst*, Lipps once more reviewed the elementary geometric forms that, according to his theory of empathy, we invest with vitality and depend on to create the architectural, ceramic, and ornamental arts (p. 401). The only explicit statement of this kind of thinking that I have been able to find among the French psychologist-aestheticians of the late nineteenth century is trivial. In a review of Léon Arnoult's *Traité d'esthétique visuelle transcendentale* [*Treatise on Transcendental Visual Aesthetics*] (1897), L. Arréat (1897, pp. 441–442) offers the following summary of some of Arnoult's ideas: art seeks permanence; the best form is geometric form; the sphere unites the elements of the greatest perfection; beauty consists "of the realization of the geometric order in the appearance of the aesthetic order" ["en la réalisation de l'ordre géométrique dans l'apparition de l'ordre esthétique"] (Arnoult, quoted in Arréat).

Although I have not been able to find a solid connection between French and German thought in the specific area of spatial form, French nineteenth-century psychology, despite its clinical and empirical emphasis, was not as remote from the German theoretical school as one might think. T. Ribot himself was the author of a book on Schopenhauer and constantly extolled the virtues of his German colleagues. He also wrote a work on German psychology, *La Psychologie allemande contemporaine (école expérimentale)* [*Contemporary German Psychology (Experimental School)*] (1885), which contains chapters on some of the figures I discuss below, such as Johann Friedrich Herbart, Hermann Lotze, and Gustav Fechner, though he gives only a sketchy summary of Wilhelm Wundt's theory of forms (pp. 272–274). Moreover, the opening flyleaf of Ribot's work advertises French translations of Wundt and Lotze. The most detailed research on the Franco-German connections in this field apparently remains to be done, but, at the very least, one can say that the French were far from ignorant of this body of material.

In my attempt to find evidence of familiarity in France with the work of the German psychologists, I have reviewed the first thirty volumes of the journal that Ribot edited from 1876 on—the *Revue philosophique de la France et de l'étranger*—which published numerous articles and reviews by and about German scholars. The first and most striking impression that one has is that the subjects of research that attracted attention in the international community of psychologists at the time are among those that are still actively investigated now: not merely cerebral localization and the nature of memory, but the aphasias, preverbal thought, inner language, subvocalization, the amusias, lucid dreaming, the differences between color and luminance, synaesthesia, the brain's revision of "raw" perception, the nature of attention, as well as the dreams of animals and the psychology of lesser organisms. The second impression one has is that there is an overwhelming preoccupation with the nature of space. Under the onslaught of the new geometries of Gauss, Riemann, and Lobachevsky, an onslaught that, in the nineteenth century, caused an upheaval in the understanding of space, the Kantian assertion that space is an a priori form in perception gradually lost support; hardly an issue of the *Revue philosophique* lacks its article on geometry, on space-time, or on the idea of dimensionality. It is easy to see how the urge to push beyond three dimensions would have fed into cubism, and how the dissatisfaction with the older notions of space and time created a breeding ground for the physics of Poincaré and Einstein. In any case, all the German psychologists and physiologists I mention above, as well as almost all those I mention below, appear as both authors and subjects of commentary in numerous issues of the journal edited by Ribot.

Henri Bergson's *Essai sur les données immédiates de la conscience* [*An Essay on the Immediate Data of Consciousness*] ([1888] 1945) offers further evidence that the French were fully aware of what the German psychologists were doing. I find Bergson's case intriguing in the context of the history of cubism because it illustrates what might be called a negative influence. Bergson seems to realize that German "psychophysics," especially as exemplified by the work of Fechner, supports the view of reality that, as we understand it, subtends cubism, namely, that the mind operates with fixed, separable quanta of experience (pp. 59, 63); that time itself can be understood in terms of space (pp. 65–66); and, at least by implication, that discontinuities need not be shunned (p. 22). In a way, then, Bergson's virtual diatribe against Fechner can be read as an attack on cubism *avant la lettre*. Bergson's well-known attachment to the idea of continuity and uninterrupted flow, and his association of "grace" itself with process (pp. 22–23), represents the opposite of the German method, which culminated not only in cubism but in quantum physics as well.

Although it is not possible to anchor each aesthetic idea proposed by the German scholars in a neurological matrix, it is apparent, from the aestheticians' regular references to psychology, not to mention that quite a few of the artists and theorists were physicians or psychologists, that the aesthetic and the physiological streams of thought were closely allied in Germany. Initially, though, confining myself to the area of aesthetics, I will simply give an example of the way in which one philosopher's argument points directly toward a later artistic development. Konrad Fiedler (1971) declares flatly that there is no stable thing called "reality" that artists can imitate, even if their mission were to reproduce reality. In any case, artists are not interested in presenting that which is given, whatever that might be. Art is not "an expression of something that could exist in the absence of that expression" ["ein Ausdruck für etwas, was ohne diesem [sic] Ausdruck ein Dasein hätte"] (vol. 1, p. xlii, fn. 42). Furthermore, the content of art is nothing but form [Gestaltung] itself (vol. 1, p. xliii).

As early as 1876, then, Fiedler provided a manifesto not only for nonrepresentational art, but also for an absolute art, one that brings to mind the Russian Suprematists. Not surprisingly, historians believe that Paul Klee was familiar with Fiedler's writings. Klee

34 wrote, in 1920, "Art does not give us the visible again: it *makes* visible" ["Kunst gibt nicht das Sichtbare wieder, sondern macht sichtbar."] (quoted by Gottfried Boehm in his introduction to Fiedler, 1971, vol. 1, p. xlv). The more complex problem of reconciling the abstract with the real, as in Cézanne, is adumbrated in Fiedler's assertion that art is the production of reality (vol 1., p. xliv).

In the development of perspectives on abstract art during the nineteenth century, it is possible to distinguish two views: 1) the belief that basic geometrical forms are the essential constituents of the visual arts, and 2) the more psychologically-oriented opinion that we respond to these basic forms because we experience them as symbolic of our own bodily attitudes and needs, as well as of the feelings associated with those attitudes and needs. The first of these ideas is, obviously, connected to movements such as cubism and its later offshoots. Its theoretical underpinnings, as I mention above, may be traced all the way from Plato's interest in the five possible geometrical solids and Plotinus' musings on symmetry and unity, through the nineteenth century, especially in German aestheticians such as Herbart, Fechner (who was also a physician and physicist), Robert Zimmermann, Adolf Goeller, and Cornelius Gurlitt.

The second view, derived from sources at least as ancient as Vitruvius' *Body of Architecture* (1931–1934; see also Rowland, 2006, p. 8, on anthropomorphism in Renaissance architecture), is closer to our immediate concern; it is associated with the attempt (especially prominent in the work of Wundt) to find a specific physiological basis for both behavior and subjective experience.

An introduction to this subject can be found in the anthology by Henry Francis Mallgrave and Eleftherios Ikonomou titled *Empathy, Form, and Space: Problems in German Aesthetics 1873–1893* (1994). (The following page numbers refer to this book.) One might begin with the quotation from Schopenhauer: "What is imagination? A very complicated physiological occurrence in an animal's brain, whose result is the consciousness of a picture or image at that very spot" (p. 10). As for symbolic form (for which Heinrich Wölfflin was later to propose a neurological correlate), Herder had already argued, in 1800, that forms are inherently anthropomorphic, i.e., that they have expressive bodily values (p. 18). In 1873, Robert Vischer emphasized the sympathy (i.e., identification) or discord of the body and its systems, especially the visual system, with what it perceives (pp. 22–23). According to Vischer, we have the "wonderful ability to project and incorporate our own physical form into an objective form" (p. 154). Writing in 1868, Lotze (also a physician) claimed that even two straight lines necessarily convey a symbolic meaning (p. 20). Vischer again: we have "an unconscious need for a surrogate for our body-ego . . . on the basis of simple nerve sensations"; every phenomenon "provokes a related idea of the self in sensory or motor form"; the phenomenon "becomes an analogy for my own structure"; one moves with a landscape, "guided by kinesthetic imagination, . . . mediated by the reflex stimuli of sensitized nerves" (p. 101). Here, again, I have not been able to discover much work on the aesthetics of empathy by the French contributors to the *Revue philosophique de la France et de l'étranger* during the years relevant to this discussion, with the exception of a debate in February, March, and June 1890 between Georges Sorel and Charles Henry over the merit of the latter's concept of *dynamogénie*, which seems to follow the familiar pattern of explaining our positive or negative response to shapes as a function of muscular stresses.

For Theodor Vischer, architecture orchestrates forces and forms that symbolize emotions and meanings (p. 19). But, for Heinrich Wölfflin, the discussion of self-projection into abstract forms, or of animation of abstract forms, lacked precision, and he preferred a more objective description of the process: "Instead of an inexplicable 'self-projection,' we might perhaps imagine that the optic nerve impulse directly stimulates the motor nerves, which cause specific muscles to contract" (p. 155). To support his opinion, Wölfflin turned to "the recently championed theory that the comprehension of human expression is mediated by a sympathetic response" (p. 155).

Here we find ourselves, of course, in the thick of the concept that dominates this field, but that I had not yet named: the rich and complex concept of *Einfühlung* [empathy], a term usually thought to have been coined by Vischer in 1873, but present or implied throughout the nineteenth century (p. 71, fn. 57). There is an automatic transfer of feeling from object to subject—we feel what we see. In a sense, we are never even entirely ourselves—we are our percepts (see Chapter 1, Section 3, above; Vischer [p. 25] seems to have been aware of this perplexing difficulty). The psychologist whose writings are most often associated with the theory of empathy is Theodor Lipps. For nineteenth-century psychologists, empathy was not a feeling confined to relations with other humans, but involved our response to our surroundings in general. Lipps explored both possibilities in two magisterial essays: the above-mentioned "Aesthetische Faktoren der Raumanschauung" and "Das Wissen von fremden Ichen" ["The Knowledge of other Selves"] (1907). The first is an incredibly detailed (eighty-nine pages, in large format) study of basic shapes, lines, and stresses, with numerous stark, unadorned drawings to drive home the point. It is hard to imagine that Lipps' thinking would not have contributed to the development of cubism and the other forms of abstract art.

The second essay is an equally extended consideration of the puzzling fact that we seem to understand the expressions on other people's faces automatically. After going through all the plausible explanations that he can think of, and finding all of them unsatisfactory, Lipps seemed to resign himself to the conclusion that empathy with others is an irreducible fact of psychology, simply a given. (For an exhaustive study of simulation theory, an outgrowth of empathy theory, see Goldman, 2006.) It was not until the 1990s that major progress could be made in this area, with the discovery of the now-famous mirror neurons. These neurons enable us to recognize other people's expressions; they lie in or near areas that are normally aroused when we see emotion in others' faces.

There remains one other historical connection to be worked out: the sources of Freudian thinking, again in the brew of mid-nineteenth-century German aesthetics. Once one accepts the idea that every form embodies some human desire or impulse, that we are constantly anthropomorphizing our surroundings, it is only a short step to the kind of somatic symbolism that Freud practiced. Actually, the chronology of this development is not straightforward. The notion of symbolic form was everywhere at the time. One version of it appears in Baudelaire's "Correspondances"; it even lurks behind the Duchess' remark to Alice (in Wonderland) that "Everything's got a moral, if only you can find it" (Carroll, [1865] 1960, p. 120).

Out of this atmosphere (or, perhaps, fog), a remarkable work was precipitated: Karl Albert Scherner's *Das Leben des Traums* [*The Life of Dreams*] (1861). Scherner has by far the most justified claim to being Freud's major forerunner, taking what was primarily an aesthetic theory and making it the basis of his dream psychology. Most of the images in dreams are understood as symbols of parts of the body, bodily states, or bodily desires; the role of the sexual apparatus and of sexual impulses in this symbolic process is worked out in particularly explicit detail. Vischer picked up Scherner's contribution in 1873 and used it to reinforce his own aesthetics (in Mallgrave and Ikonomou, p. 100). Two years later, in 1875, another of what one might call the "projectionist" aestheticians, Johannes Volkelt, wrote at length about Scherner in *Die Traum-Phantasie* [*Dream Fantasy*], attempting to draw attention to what he considered Scherner's crucial contribution to psychology. (In his 1886 *Prolegomena to a Psychology of Architecture*, Wölfflin tried to find a neurological grounding for Volkelt's theory of symbolic projection.) At some point between 1875 and 1899, the year *Die Traumdeutung* (*The Interpretation of Dreams*) was published, Freud found his way to Scherner's work, apparently via Volkelt, whom Freud cites extensively. Freud owned two copies of Scherner's *The Life of Dreams*, both of which he underlined heavily.

I first drew attention to Freud's debt to Scherner in "Freud before Freud: K. A. Scherner (1825–1889)" (1990). Scherner is an interesting but mysterious figure, heavily buried under the sands of intellectual history. He never wrote the promised second volume of his work on dreams; in fact, the only other book of interest that he produced was a guide to the Tatra Mountains. His health, both mental and physical, appears to have been precarious. Many other factors contributed to his obscurity; members of his university (Breslau) told me that among these factors were his poverty, his choice of neighborhood, and his membership in the dissident group of the "old believers," who were presumably contemned by Scherner's peers, especially at a university under ecclesiastical control. That university has certainly made no attempt to commemorate his presence there, though his biography would be well worth a full-length study. He was also eclipsed by more famous philosophers such as Wilhelm Dilthey and Lipps himself, who followed him at Breslau.

Not least among those who were responsible for the failure of his reputation, however, was Freud. Freud referred to Scherner frequently and praised him highly; at the same time, he was extremely careful to avoid citing precisely those passages in *The Life of Dreams* that clearly anticipated his own ideas. In any case, it is interesting to see how Freud—who began, of course, as a neurologist—allowed himself to be seduced by the psychologized remnant of an aesthetic theory in Scherner and subsequently changed course in midcareer, and how some of his major ideas derive from that theory. It is also interesting to realize that *Die Traumdeutung*, a work that came to be regarded as the founding text of a revolution in psychology, actually represented a retrograde movement, the revival of what was, by 1899, a fading paradigm in aesthetics.

Of course, phrases such as "symbolic form," associated primarily with the names of Ernst Cassirer and Suzanne Langer, also had a strong career in the twentieth century. Aestheticians such as Rudolph Arnheim and Matila Ghyka may be considered disciples of Lipps; indeed, Ghyka (1952) is simply Lipps taken to extremes; and as late as 2004 we still find advocates of the significance to be found in elementary forms (and, indeed, of a universal grammar of forms), such as Peter D. Stebbing (see also D. W. Zaidel, 2005, pp. 62, 240). It is interesting to note that Stebbing, in his "A Universal Grammar for Visual Composition?" (2004), finds it necessary to invoke the authority of Humboldt. And a famous experiment conducted at Smith College in 1944 showed that, when triangles, a square, and a circle are set in motion, most viewers promptly endow them with human characteristics and an accompanying narrative (Heider and Simmel, 1944). Nevertheless, the large and concentrated school of thought that focused on the concepts with which these authors have worked was a phenomenon of the nineteenth century.

In any case, the above historical section is meant to show the connection between theories of art and psychology during that earlier period. I will now offer a brief overview of some psychological, or rather, physiological approaches to art in our own time. My concern in this second part of my chapter will not be to follow an historical development, but rather to ask what light the neurological tools and concepts now available to us cast on the work of art: Where do they provide access to aspects of art previously understood either intuitively or not at all, and where do they fail to contribute anything, or promise more than they can provide?

2. NEUROLOGICAL THEORIES AND THE VISUAL ARTS

Neuroscience has contributed widely to contemporary ideas about art and art education. Semir Zeki, who specializes in the visual arts, is one of the most frequently cited neuroaestheticians. His best-known work, supplemented by numerous articles and lectures, is *Inner Vision: An Exploration of Art and the Brain* (1999). By now, one might expect that visual art would lend itself readily to neurological investigation: enough is understood about the visual system to enable us to correlate, say, the op artists' work almost point by point with well-recognized processes in the visual cortex. Neuroscientists know what causes many optical illusions: for example, they know how a static image can cause the impression of movement or a sense of tension. In *Vision and Art: The Biology of Seeing* (2002), Margaret Livingstone painstakingly distinguishes the effects of luminance from the effects of color, and she uses this distinction to account for many features of Impressionist painting. (On the neurology of the "what" vs. the "where" systems, so important for Livingstone's analysis of the disorienting effect produced by some of Monet's paintings, see Heilman, 2005, p. 67.) Obviously, there is a connection between artistic style and the brain's behavior: the aphasiologist and aesthetician Jason Brown (1999, p. 146) mentions a Bulgarian artist, an exponent of social realism who adopted a bold experimental style after suffering a stroke, and Dr. Jean-Marie Annoni has closely studied the ways in which strokes affect artists' styles (see B. Carey, 2005). Much of Dahlia Zaidel's *Neuropsychology of Art: Neurological, Cognitive, and Evolutionary Perspectives* (2005) is devoted to the examination of such cases (see also Chapter 1, Section 2a, above).

It would seem, then, that there is almost a ready-made vocabulary for talking about the visual arts and neurology: use a certain color in the middle and another color in the surround, and you get a certain effect; emphasize line rather than color, and you get another effect; use motion without color, and you get yet another effect. If you add colors to a Calder mobile, the brain will not respond as well to the mobile's motion.

Nor has the systematic attempt to classify compositional strategies in music been abandoned. Music theorists have been trying to do something of the sort ever since Johann Mattheson published his influential dictionary of musical effects in 1739; in fact, long before. It is easy enough to specify what combinations of pitch and tempo induce feelings of relaxation, of anticipation, of urgency, of closure, and so forth. For obvious reasons, the rules of classical composition could be worked out only on empirical, not on neurological, grounds. (In this respect, they resemble the architectural rules that Rudolf Arnheim [1977], following Lipps, as I have mentioned, attempted to establish.) Recently, though, David Huron (2006), following Leonard Meyer, has tried to analyze the feeling of musical expectation in terms of cognitive science and neurology. Still, a neurology of musical effects remains elusive.

In the visual arts, however, such specification is actually attainable, as visual functions can be mapped clearly on the neural

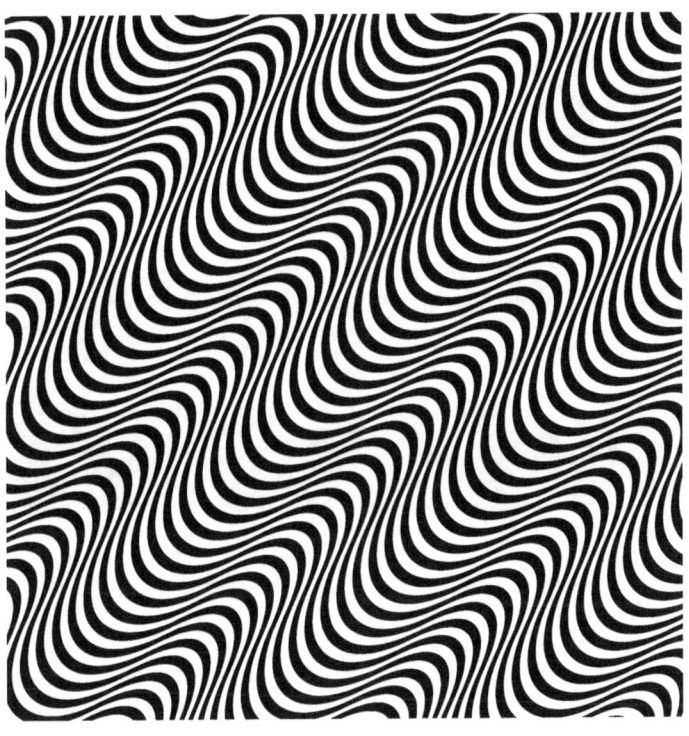

FIGURE 1A. *"Undulating Lines." Image produced
by David A. Moore, graphic design artist, Buffalo, NY.*

structure of the brain. But, in some odd sense, the very ease with
which the neurological analysis of artistic processes can be applied
to, say, painting, makes the enterprise seem futile. Even the most
sophisticated book on art and neurology starts to look like Keith
Kay's *Little Giant Book of Optical Illusions* (1991), and Livingstone's
wonderful full-color plates from Monet, Matisse, and Mondrian,
not to mention the trick pictures by Akiyoshi Kitaoka ("Rotating
Snakes" [Figure 3b]) and David A. Moore ("Undulating Lines" [Fig-
ure 1a]), or Bridget Riley's *Fall*, or Roy Lichtenstein's dim cathedral
(likewise reproduced by Livingstone), begin to look like illustrations for an ophthalmology textbook. Incidentally, the disconcert-

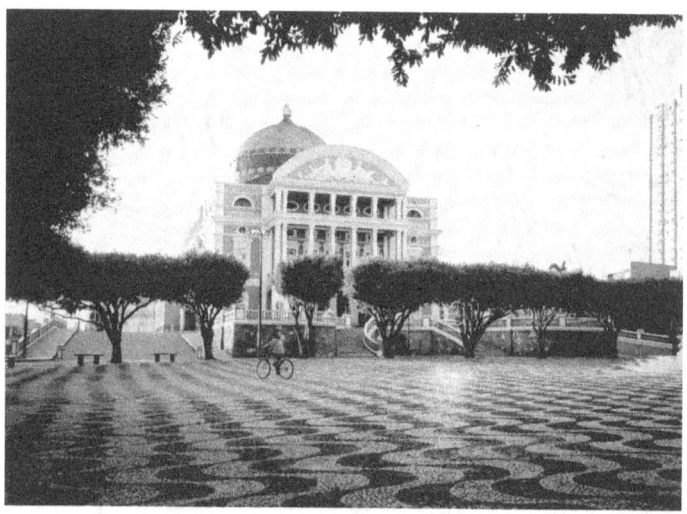

FIGURE 1B. *Photograph of the square in front of the opera house
in Manaos, Brazil, taken by Ann Colley, December 25, 2005.*

ing sense of depth and motion on a flat surface that one has when
viewing Figures 3b and 1a can be experienced on a large scale when
one confronts the design in the pavement of the square before the
opera house in Manaos, Brazil (Figure 1b).

The problem with the neurological analysis of effects in art is easy
to find: it is that vision is not in itself aesthetic, any more than any
of the other senses is in itself aesthetic, certainly not in the modern sense of the word "aesthetic," i.e., "having to do with art." (As
I have mentioned, the early eighteenth-century usage would have
been more like "having to do with sensation.") Zeki (1999, p. 210)
claims that Cézanne and Monet repeatedly painted the same scene
in their search for constancy under different conditions and "thus
mimic[ked] unknowingly the function of the visual brain." Cézanne and Mondrian, in seeking "the essential constituents of all
forms," emphasized "those very stimuli which are the most effective for activating single cells in the brain" (p. 209), and "the properties of these cells are, to an extent, the pre-existing 'idea' within

us" (p. 124). But if sensation as such is the pre-existing idea, why bother to strive for art? It is a perfectly respectable goal in art to search for "essences," but not when they turn out to be merely the ordinary functioning of the senses. Zeki seems to confuse the earlier meaning of "aesthetic" with its present use. Similar objections to Zeki's argument have been raised by John Carey in *What Good are the Arts?* (2005, pp. 87–88) as well as by W. P. Seeley (2005).

Although Zeki claims to have rejected the Platonic notion that ideas exist outside their various embodiments, he seems to have replaced them, at least in this book, with the internal ideal of the single cell, his version of the divine essence. Time after time Zeki invokes the functioning of the brain as the standard for what art should try to achieve, apparently never realizing that if art did achieve that standard, all we would end up with is the perfection of ordinary perception rather than a work of art. If art mimics what the brain usually does, it does so only in order to manipulate that effect. It does not do what the brain would be doing anyway, of its own accord. Art emphasizes "inappropriately." Art does not duplicate the natural percept; it cancels the natural percept and replaces it with another. In a melody, for instance, what it cancels is the note as it existed *outside* the melody; in a song, the word as it existed *outside* the song. Similarly, in rhyme, meter, or other poetic devices, it works at cross-purposes to the perceptual brain, continually defying our prosaic expectations and linguistic norms. Zeki, ever the crypto-Platonist, says that art "makes us independent of the single and fortuitous view" (p. 68); I would say that the work of art does, indeed, liberate us from the fortuitous view, but only by creating a view that is systematically at variance with the one provided by ordinary perception.

Again and again, Zeki tries to account for artistic effects in neurological terms. Of course, cells with a preference for a particular orientation will react vigorously to a Kasimir Malevich, an Olga Rozanova, or a Barnett Newman; but they will also react vigorously to any lines of that orientation that appear in nature. The fact that these cells respond to such lines tells us nothing about the work

44   of art within which the lines appear. To be sure, a certain kind of aesthetic experience would be unavailable to us if we lacked these cells (p. 116), but so would an *un*aesthetic experience. If verticals and horizontals are easiest for us to see (p. 115), and if there are no cells that respond to curves as such (an opinion, I gather, that Dr. Moshe Bar does not share), does this really (as Mondrian would have it) make curves less aesthetic than verticals? One might almost take Warhol's Campbell's soup cans as a parody of the kind of aesthetics toward which Zeki pushes us: it is "You get what you see," reduced to an essence; but that essence coincides with what is already there: the can.

In his more recent work, as, for instance, in his March 2005 presentation at Harvard, Zeki has been emphasizing the unfinished quality of great art (Michelangelo, Wagner, etc.) as a reflection of the artist's pursuit of an unattainable ideal. (The superiority of the unfinished to the finished was a popular doctrine among the German Romantics, among them Friedrich Schlegel, and it was revived by John Ruskin.) But the main point is that here, again, a Platonic principle, i.e., that no incarnate form can do justice to the Idea, is being invoked by Zeki, without being named. It is an ancient opinion in aesthetics, and hardly requires the support of neuroscience to be deployed. I have also observed (and John Carey [2005, p. 84] has had a similar thought), that the goal of such an artistic effort would be inaccessible to a neurological description, being, by definition, unavailable for any sensory incarnation.

Perhaps the familiar Platonic project is not really the best on which to frame a theory of the plastic arts, since it moves toward the rejection of art, or at least toward the jettisoning of art on the way to some other sort of perfection. The pleasure or interest of a work of art always starts with, and depends on, its embodiment. Why assume that Monet was seeking a single, ideal cathedral behind his multiple views of the one at Rouen? After all, there is an alternative aesthetic, the one that would value each individual view precisely for its own sake, its "quiddity," its Scotist "isness." We revel in the realization that a single form can yield so many realities, each

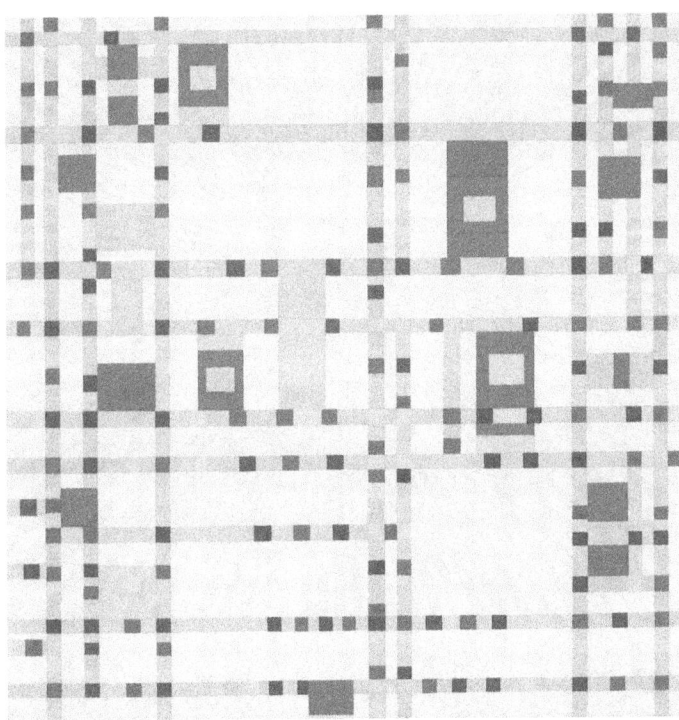

FIGURE 2. *Piet Mondrian,* Broadway Boogie Woogie, *1942–1943.*
© HCR International. Oil on canvas, 127 × 127 cm. The Museum of
Modern Art, New York, NY. Digital image © The Museum of Modern
Art/Licensed by SCALA/Art Resource, NY.

complete in itself. The full satisfaction that we take in any one of
the paintings should not be squandered in the search for something
"behind" it, some absolute building without qualities, against which
it is to be measured. Its perfection in the moment, brief in the act
of perception but eternal in its capture, is what we value. It is what
I have elsewhere called a "then-forever" (1987a, pp. 32–33).

When it comes to the criticism of specific works, even a more
focused scientific approach than Zeki's does not advance us very
far. Take what might be called the poster child of the neurology
and art field, Mondrian's "Broadway Boogie Woogie" (Figure 2),

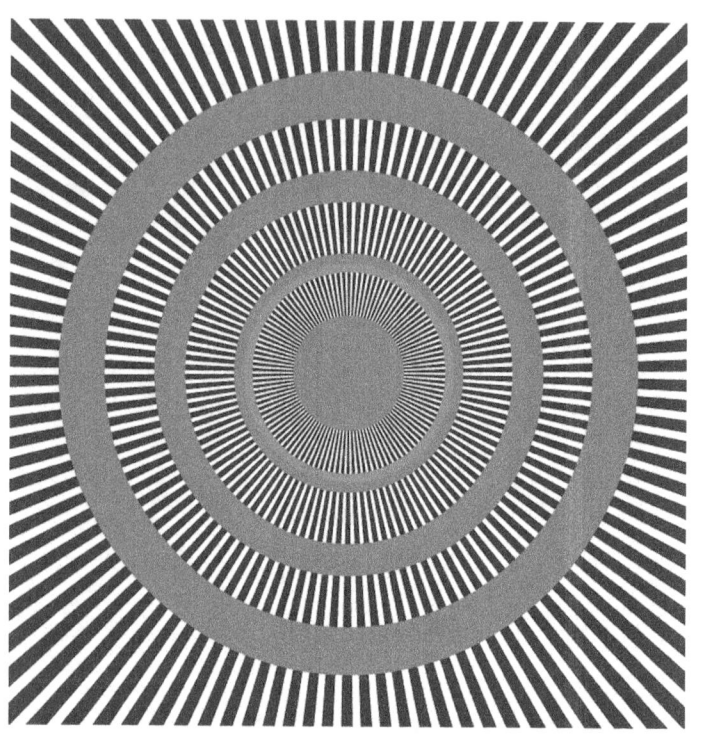

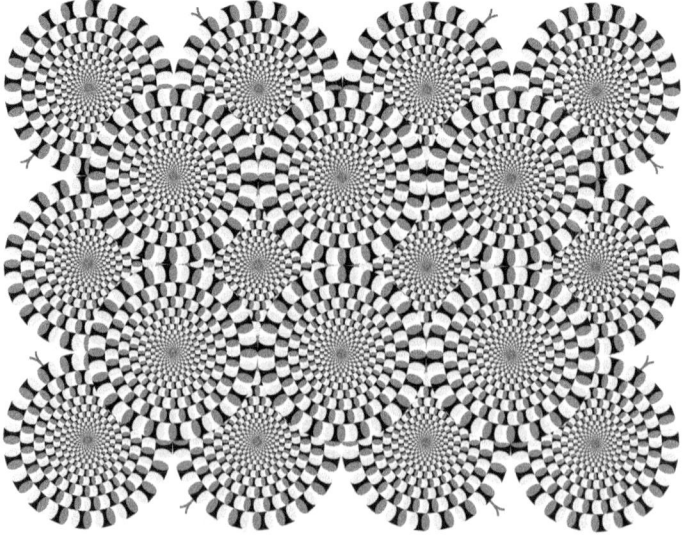

FIGURE 3. *(a) Isia Leviant, "Enigma." Permission granted by daughter of the artist, Catherine Couvreur. (b) "Rotating Snakes." Reproduced with permission of Prof. Akiyoshi Kitaoka, Ritsumeikan University, Kyoto, Japan.*

which is reproduced in almost every study of this sort. Presumably the painting suggests a city moving to an irregular beat. Margaret Livingstone (2002, p. 157) offers an elaborate theory in which she claims that the picture produces a jittery effect because "the yellow and grey squares are indeed very close to equiluminant with the off-white background." (On the other hand, the art critic Clement Greenberg, notoriously, saw colors in the picture that were not even there [see Jones, 2006, p. 25].) For me, perhaps because of my poor eyesight, this is not at all where the interest or indeed the effect of the picture lies: it is rather in the incessant frustrations that this fairly large painting engenders, when forms that one expects to be repeated exactly turn out to be noticeably different in size, shape, or placement each time they reappear, while the whole thing is balanced on an inappropriately small red rectangle at the bottom center. There are many things that keep the eye moving in this picture, dwelling now on this and now on that—as they do, indeed, in many a more traditionally conceived painting. Actually, if one looks carefully at the original, one finds it to be very painterly in texture, almost in defiance of the geometric forms; this is, in fact, one of the most interesting contrasts that the work produces. But the point I am trying to make is that no formula in optics can pin one down to a single way of perceiving a work, and that traditional subjective strategies are ineradicable from the contemplation of even the most optically oriented work of art. Even with a "trick" painting, such as Isia Leviant's "Enigma" (Figure 3a) or Akiyoshi Kitaoka's "Rotating Snakes" (Figure 3b), in which a spinning effect is produced by a stationary image, the question remains, as I have asked in my first chapter, what does the spinning do for the picture? (For one neurological explanation of the process by which the illusion of rotation can be produced in a stationary image, see Haseltine 2003.) Is it there for its own sake? Is it merely a device, or does it contribute to some purpose beyond its motion? Does one end up liking the picture, or finding it important in some way? If so, why? Is it merely because it spins? One could make innumerable other pictures that produce the same effect. Must they all be

48 equally good? On what grounds would we discriminate among them, in terms of quality?

Neuroscience may illuminate certain processes associated with aesthetics, such as synaesthesia, a newly revived area of interest; it may certainly provide a new context for the discussion of a poem such as Rimbaud's "Voyelles" (1967), in which a different color is assigned to each of the vowels, helping to explain the clang effect of certain syllables as they evoke a particular hue; but it will not help us to decide whether we like the poem, or how much.

To return to Margaret Livingstone's *Vision and Art*: at times, Livingstone undermines her own strategy. She goes so far as to say, "Indeed, I find enlarged magazine images to have colors that are just as visually interesting and just as lively as Post-Impressionist paintings" (p. 176). In other words, because the elements of color printing follow some of the same principles as the techniques of pointillist painting, they have the same artistic value. This conclusion, I believe, almost deliberately weakens the purchase of neuro-criticism on aesthetics. Like Livingstone, Dahlia Zaidel (2005) confuses fields (in this case, pathology and art history) when she attributes the rise of Impressionism to the defective vision of certain artists (p. 73).

It remains to be asked whether, in spite of all these objections, there is some clear-cut benefit in applying optical principles to the study of the visual arts. Perhaps it would be best to approach the question from the negative side: Do the visual cortex and its related structures have certain characteristics that constrain the eye and either forbid or oblige us to respond to artworks in certain ways? Does the visual function have limitations that the artist must either submit to or try to circumvent? (Similarly, I suggest in Chapter 4, Section 8, that right hemisphere dominance in music constrains the function of words in song.)

A case of what used to be called, in a Gestalt context, "Praegnanz," Figure 4 provides a simple example of how our visual faculties constrain our perception. Ramachandran and Hirstein (1999, pp. 22–23) argue that, for neurological reasons, when we look at the image, our minds prefer to perceive parentheses rather than hour-

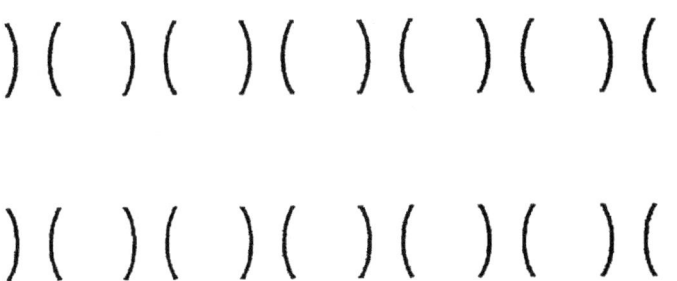

FIGURE 4. *"Gestalt grouping principles. The tokens can be grouped either on the basis of 'proximity' (which produces hourglasses), or 'closure.' The latter organisation is more stable and pleasing to the eye" (image and caption from Ramachandran and Hirstein [1999], p. 23).*

glasses; closure trumps proximity. Mary Cassatt's *Young Mother* (Figure 5) provides an artistic example of the same phenomenon. In the painting, the brown and blue areas, respectively, may be said to form parentheses, but the absence of the child's arm completing closure on our right creates a sense of instability, reinforcing the child's slightly anxious expression and the perhaps thumb-sucking gesture. Almost the same pattern is repeated in Cassatt's *Children Playing on the Beach* (Figure 6): a complete pair of parentheses appears on the left, and an incomplete one on the right, but it does not create as strong a sense of instability as does the unfinished pair in *Young Mother*; it does not generate the potential fear of falling off. The little girl in *Child in a Straw Hat* (Figure 7) has one complete set, but her tilted head and slightly unhappy look throw off the sense of security that the arms, joined at the hands, provide by holding in the straining, broad hourglass of her chest and preventing it from exploding the parentheses that try to contain it.

Frank Stella (1986, p. 6) describes even the pictorial space of the "Mona Lisa" in terms of parenthetical shapes, as a "transparent membrane" with "two beautiful, slightly bulging soap bubbles bound together." Once one subscribes to the principle of the reigning parentheses, one begins to see them everywhere. I look at a

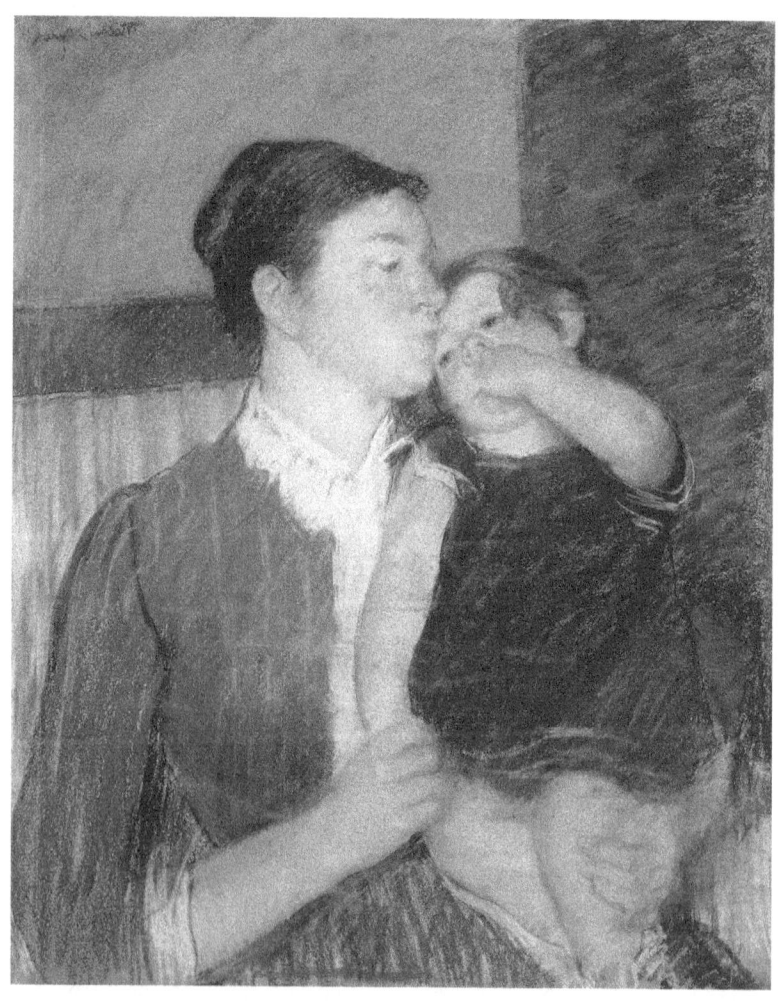

FIGURE 5. *Mary Cassatt,* Young Mother, *1888. Pastel on blue-gray wove paper faded to tan, mounted on canvas, on a strainer, 84 × 73.8 cm. The Art Institute of Chicago, Chicago, IL. Photograph © The Art Institute of Chicago.*

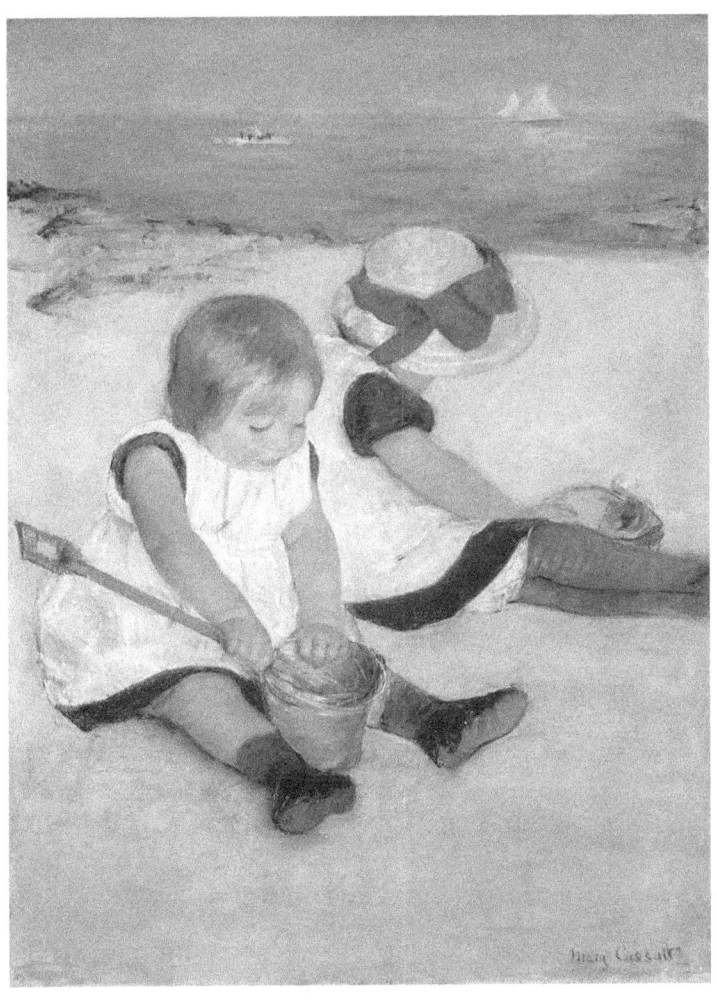

FIGURE 6. *Mary Cassatt,* Children Playing on the Beach, *1884. Oil on canvas, 97.4 × 74.2 cm. Alisa Mellon Bruce Collection, National Gallery of Art, Washington, D.C. Image courtesy of the Board of Trustees, National Gallery of Art.*

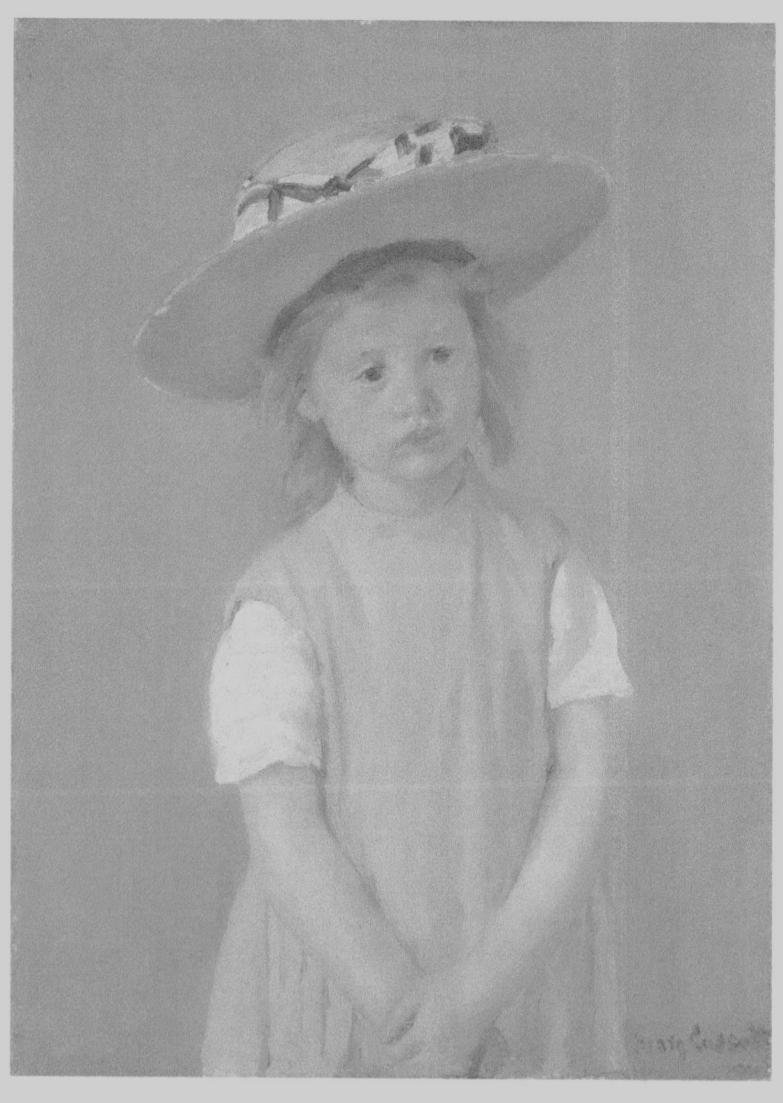

FIGURE 7. Mary Cassatt, Child in a Straw Hat, *c. 1886. Oil on canvas, 65.3 × 49.2 cm. Collection of Mr. and Mrs. Paul Mellon, National Gallery of Art, Washington, D.C. Image courtesy of the Board of Trustees, National Gallery of Art.*

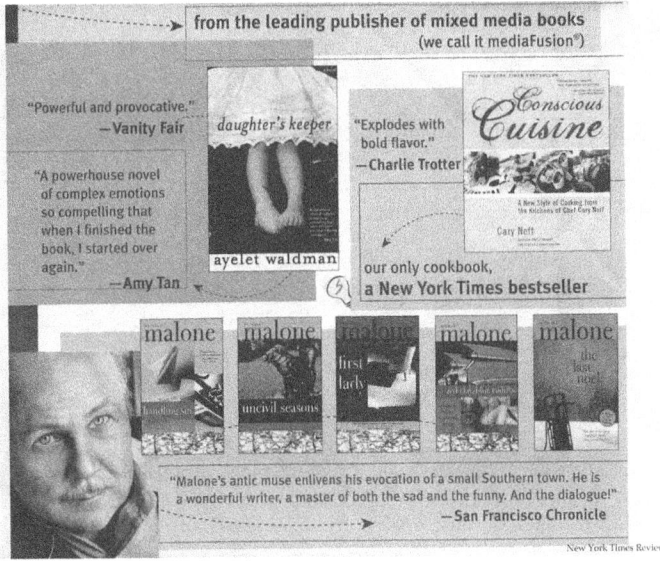

FIGURE 8. *Page of book ads in the* New York Review of Books. *Permission for incidental use has been kindly granted by the* New York Times Review *and Getty Images. The full page may be found in the* New York Review of Books, *November 20, 2003, p. 7.*

page of book ads in the *New York Review of Books* (Figure 8), and it jumps out at me from among the other ads. I look at Turner's *The Burning of the Houses of Parliament* (Figure 9), and the whole left side of the canvas appears to me to be two mirrored sets of parentheses, continued by the arches in the lower right quadrant. But, in the end, if one is honest with oneself, one must acknowledge that all of this is probably just a subjective response masquerading as a scientific observation.

Of course, it is not even clear what "looking" at a picture means: the eye is engaged in constant saccades (though perhaps a baby's vision is different), and we shift focus continually; some pictures exploit this fact (see Hogan, 2003, pp. 81–82, with references to Robert Solso, 2003; see also the quotations from Schelling, Goethe, and Ampère in Crary, 1990, pp. 99–100). On the other hand, complete

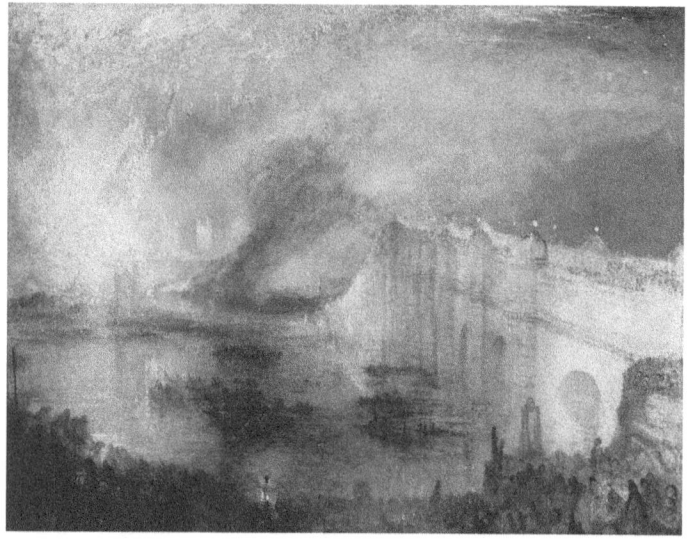

FIGURE 9. *Joseph Mallord William Turner,* The Burning of the Houses of Lords and Commons, October 16, 1834, *1834 or 1835. Oil on canvas, 92.1 × 123.2 cm. The John Howard McFadden Collection, Philadelphia Museum of Art, Philadelphia, PA.*

discontinuity of vision is a symptom of certain kinds of brain lesions (Heilman, 2005, p. 63). Maybe our vision has a built-in rhythm that alternates our awareness of figure and ground. Or does the visual field sometimes belly flop, bulge inwards? Frank Stella (1986, p. 6) divides the landscape in the "Mona Lisa" into two such bulges, a concave bulge and a convex bulge.

In any case, we never see a whole picture. When looking at one, we all suffer from what Dahlia Zaidel (2005) calls "simultanagnosia." There is a constant conflict of space vs. boundary, gravity vs. uplift, focal vs. peripheral. On occasion there is even a fluctuating awareness of the image, of the image's "meaning," and of the materials that make up the image. It is possible that Matisse, with his arabesques and with the textile artist's "abolition of boundaries" (Labrousse, 2004, p. 55), was trying to liberate us from the inappropriate and impossible obligation of seeing a picture steadily and seeing it whole, and from the guilt and dissatisfaction with which that effort leaves us (cf. Labrousse, 2004, p. 57). Perhaps, in certain ways, remembering a picture gives us a more satisfactory experience than the act of looking at it does.

I have puzzled over the difference between the way we apprehend a building and the way we apprehend a picture. For me, at least, the initial impression of a building is that of a whole. What is more, the response, favorable or unfavorable, is instantaneous. Only afterward do I think of what I like or do not like about it, about the nature of its structure, its colors, or its ornamentation. Perhaps we make an instant decision that some large object in our environment is supportive of or inimical to our needs, and only later allow ourselves the luxury of asking why (cf. Hogan, 2003, p. 174), but I do not really find this explanation satisfactory. I do believe, though, that the appearance of a building either hurts us or helps us. Naturally, we look at a building and a picture differently: a building creates its own boundaries, whereas a picture is always actually or implicitly framed. A building asserts itself toward us in a way in which a picture, no matter how large, does not. We need the building, whereas the picture needs us.

Before concluding, I wish to discuss some philosophical considerations which, I would like to show, are related to certain neurological facts. On the whole, the theory of the image during the twentieth century tended toward iconoclasm (see Massey, 1987a, p. 10, where "Warburg" should read "Würzburg"). Philippe Lacoue-Labarthe expresses a characteristic attitude in *La Fiction du politique* [*Heidegger, Art, and Politics*](1987, pp. 126–127). Lacoue-Labarthe claims that the attempt to represent something is destructive either of the self or of nature: either one abandons self in the act of representation, or one rejects nature, having found it inadequate, and replaces it with one's representation of it. In the process of representation one is either too aggressive or too passive. (It does not seem to occur to Lacoue-Labarthe that a representation may be an expression of admiration for the object, a positive engagement with nature that is like an act of love, in which one gives and receives, is at once active and passive.) Jean-Luc Nancy (2003, pp. 49–50) also speaks of the violence associated with imposing contours or definitions on anything, or with trying to create the illusion of unity, of which the image is a prime example. On the one hand, the image is often associated with violence, with seduction (Nancy, 2003, p. 26), with the illicit in one form or another. On the other hand, the transgressive quality of the image has often elicited an even more transgressive reaction: one has only to think of the February 2006 Danish cartoons of Mohammed and the later riots to be reminded of the power of iconoclasm and of the concentrated violence that has been unleashed repeatedly in its name, throughout history, beginning, if one chooses, with the destruction of the golden calf.

But the truth is that the image is ineradicable because, as some philosophers have surmised ever since Aristotle (1941), the impulse to represent is ineradicable. Arnold H. Modell (2003, pp. 183–192) suggests that the mirror neurons may have been involved in the origins of language and of the metaphoric process. J. R. Skoyles (1998) attributes the representation of motion in Greek art after the Archaic period to an instinctive realization of a kinetic bond between the viewer and the art object, similarly attributable to the activity of the mirror neurons; in fact, Molnar-Szakacs and Overy (2006) seem to find a role for the mirror neurons in almost every situation!

In spite, then, of Tom Huhn's (2004) subtle argument (if I understand it) that, even in eighteenth-century aesthetics, the mimetic principle persisted primarily as an occulted awareness of social experience; in spite of Jonathan Crary's (1990, p. 91) observation that, as early as 1833, Johannes Mueller had "unsparingly laid bare" the "referential illusion"; in spite of Konrad Fiedler's 1876 manifesto for nonrepresentational art; in spite of the attempt by Lacoue-Labarthe and others, in a well-known book (Agacinski et al., 1975), to "disarticulate" mimesis—representation in art goes on. Sometimes there is a representational subtext in an apparently abstract artwork; more to the point, it is still a representational painter, Lucian Freud, who is now regarded by many as the most important contemporary artist. Some might consider this situation deplorable, but what I am suggesting is that, for better or for worse, representation is not something that is going to go away.

Alain Badiou (2007) argues that the hostility to representation that has been ubiquitous since the early nineteenth century is a specific historical phenomenon that may fade as the *Zeitgeist* changes. Badiou attributes the hostility to representation not to the usual culprits such as the rise of photography or the emergence of non-Euclidean geometry, but to a much broader cause, namely, the rage for the absolute: a pervasive finalism that regards representation as merely a step toward its own eradication. I quote Alberto Toscano's translation of Badiou: "Modern art brings the relative universe of representation to ruin," and "In art, it is thought that by putting an end to the relativity of imitations and representations absolute art will be attained" (p. 36). Such a stripping away, or purging, of all that is relative—i.e., of representation—brings us finally to the real, but that "real" turns out to be nothingness itself: death (p. 54) or, at most, a gap (p. 56). The implication of Badiou's reasoning is that a century less obsessed with finalism would find representation less threatening.

But whatever the threat of or to representation, the point is that representation is not something artificial. It is not merely something that is made up according to certain conventions. For those who would like to get rid of representation, the problem, as the nineteenth-century aestheticians who developed the idea of empathy made clear, is that the act of perception is itself partly an act of imitation; at the very least, it is a negotiation, sometimes a struggle, between the endogenous and the mimetic. (A literary footnote: could Christabel's "forced unconscious sympathy" in Coleridge's poem [(1816) 1957, line 609] owe something to Adam Smith's idea of unconscious mimicry [see Goldman, 2006, p. 276]? Dugald Stewart also spoke of involuntary "sympathetic imitation," and T. H. Huxley later was to characterize imitation as a pure reflex [Amigoni, 2007, pp. 21–22].) Empathy brings us into relation with

the world at large as well as with the other living beings that we
have to understand. Obviously, we have to know how to introject
the expressions of other people in order to interpret them. (Jason
Mitchell et al. [2005] localizes parts of this process in the medial
area of the prefrontal cortex.) The more general "empathy neurons" appear to have been recently localized (Tankersley, Stowe,
and Huettel, 2007), and it might not be going too far to say that the
mirror neurons may supply what might be called a native foundation for mimetic art, especially for that art that deals with living
forms, and with the human in particular. (For a pointed rebuttal,
see Gopnik, 2007, as well as Saxe, 2005.) Their existence and their
function could help to account for the tenacity and the perpetual
revival of the impulse to imitate.

3

At the Limits of Language

1. APHASIA, DREAM, AND POETRY

In 2001, Ronald Schleifer published a groundbreaking essay, "The Poetics of Tourette Syndrome: Language, Neurobiology, and Poetry." In that essay, Schleifer set out to demonstrate the affinity between, on the one hand, the language used by patients suffering from a specific neurological disorder and, on the other hand, some forms of poetic language. I would like to take Schleifer's project a step further by exploring other situations where language use falls into unusual patterns; eventually, I will also try to demonstrate the possible relevance of these other phenomena for metaphor and for poetry.

My intent in the first section of this chapter is both more specific and more general than Schleifer's. In Subsections a–d, I offer the specific, or rather, the technical conjecture that, during sleep, language areas in the brain are partially dissociated from the areas that process images and action. I suggest that this condition calls for study by means of brain imaging techniques. In Subsections e–j, I make the second, far more general suggestion that dreamlike language represents an undertow in all our expression. The relation of symbol to referent and/or concept to image is inherently unstable, and dream language exists in, or exploits, that ambiguous space. I do not quite wish to assert, with Rodolfo Llinás and Denis Paré (1991), that consciousness itself is just a partially suppressed dream, but that dream language is a kind of fallback, or default setting, for our "executive" selves.

a. Bizarreness and Dream Language

I will begin with dream language. My point of departure will be the fact that the language one hears, or overhears, in dreams is sometimes wildly at variance with the action and imagery of the dreams, dissociated from everything else that is going on. Language in dreams can be as "bizarre" as the images that it accompanies, and may resemble some forms of aphasia. This phenomenon is more frequently observed during transitional sleep states (such as hypnopompic sleep) than during other sleep phases, but it is by no means confined to these transitional stages. (In this chapter, "dream" will include dreams that occur both during such transitional states, during NREM, and during REM sleep, since these distinctions are not material to my purpose.) The qualitative differences among dreams during the various stages of sleep are, of course, still the subject of intense debate (see, for example, Solms, 2000, p. 845, and commentary by Hunt, 2000, pp. 954–955).

I take dream to be, historically, prelinguistic: REM sleep, with dreamlike behavior, can be found in almost all mammals. Although David Foulkes (1999) maintains that only postinfantile humans, with their special linguistic/cognitive abilities, really dream, I find his thesis, cogently argued though it be, finally unconvincing, and it now seems to be fairly well established that rats have visual dreams (Wade, 2006). It also seems clear to me that small children dream. To be sure, it cannot be denied that in humans almost all mental states and processes, sleeping or waking, are touched by language, and that many dreams display narrative elements that also suggest a linguistic process. Once dreaming became human, it had to have a linguistic component in its very foundation. Nevertheless, language can work in many different ways, and it may be dominant or subordinate to other mental functions in different situations. Given the indications that dream developed before language appeared on the evolutionary or the mental screen, it is not surprising to find that language acts differently when it operates in the dream domain, an independent system for which (unlike the conscious mind) there is no survival benefit in giving priority to language. Perhaps dream tells us a little about what the mind was like before it came to be dominated by language; it may afford us a glimpse into our prehistory. Besides, the limitations that the condition of sleep usually imposes on language are not entirely to language's disadvantage. Although from one point of view it may appear defective, from another perspective dream language can be seen as often successful in expressing inchoate, blurred, or overlapping thought-moods that ordinary, syntactically complete, monosemantic language could not even begin to deal with. (Blechner [2001, pp. 17, 25–26] has excellent comments on this theme; I thank Patricia Kilroe for the reference.) Dreams even have a way of jogging us into the uncomfortable awareness that thought and word may be incompatible, and that our regular language cannot come into being until it has set thought aside.

b. Aphasia and Dream Language

There is another possible reason why dream is dominated by images rather than by language: perhaps the brain drops into aphasic patterns during much of our sleep, so that, even if the organism wished to express itself in normal language, it would be prevented from doing so. In other words, perhaps we dream mainly in images not for psychological reasons (as Freudians, for instance, might maintain) but for physiological reasons. If the brain resumes some of its earlier patterns of behavior during sleep, channels only recently specialized for speech could revert to their previous function, and their use for speech could be obstructed or abandoned. (The late Elizabeth Bates of San Diego argued that "language is overlaid on areas that do sensorimotor work" [see Corballis, 1999, p. 144, and also Darwin, [1871] 1902, vol. 2, p. 735].) Some "runs" of dream language are similar to "fluent aphasia," and, as Kraepelin long ago implied, might lead one to want to look for alterations in the function of Wernicke's area during dream (Heynick, 1993, p. 141). (I will return later to the relation between dream language and aphasia.

To follow up on the discoveries of Toronto's Prathima Shammi and Donald Stuss, one might also want to investigate the possibility that physiological constraints could help to account for the paucity of humorous material in dreams [see E. Johnson, 2002, p. 24].)

Of course, language in dream appears in many forms. Sometimes it is simply the dialogue, or monologue, of characters in the dream, one of whom may be oneself. At other times, it consists of marginalia that are not integrated into the dream; one is simply aware of its presence (see the quotation from Nabokov, below). In dream, words, as a rule, adhere to images or to actions rather than confront them. They hang around the dream rather than express it; they often seem to have observer status. Sometimes they comment on the action or pun on it; what they will not do is "tell." This sticky quality of the language, as though it clung to the surface of the action, rather than existing in a separate dimension, as any self-respecting language is supposed to do, gives it the oddly physical quality that Gerard Manley Hopkins attributes to it. The image of the word for the self, he says, "is in fact physical"; as for the word in dream, "it comes from the involuntary working of nature" (1959, p. 125). I should add that, even when language appears to be appropriate to the action in a dream, and is neither fragmentary nor incorrect, it still does not seem to work in quite the same way as it does when we are awake. Its occurrence in dream always seems specialized, as if it existed for some reason other than mere communication or reasoning.

To be sure, if dreaming preceded language, there is no reason why it should engage in ruminative thought, nor make its contents available to consciousness in discursive form. Why should dreams make sense? Nabokov (1966, p. 33) writes, "Just before falling asleep, I often become aware of a kind of one-sided conversation going on in an adjacent section of my mind, quite independently from the natural train of my thoughts. It is a neutral, detached, anonymous voice, which I catch saying words of no importance to me whatsoever." (Ian Oswald [1962, pp. 106–115] gives numerous other examples from hypnagogic sleep.)

c. Language and the Other

One of the great disadvantages of conscious language is that it is never entirely one's own. Language is just that part of thought that relates to others. To quote myself (1976, pp. 30–31),

> in dream we speak our own language; when we wake, the language of others. Surrendering our sleep is difficult not so much because it requires us to interrupt our fantasies as because it forces us to assume the language of others, to begin to talk . . . the catatonic soldier [in Nerval's *Aurélia*] is fortunate because his thinking is "sans le mélange de la pensée d'un autre" ("without the admixture of someone else's ideas"). This is the burden and the strain of consciousness: that we must speak a borrowed language.

John Hughlings Jackson, writing in 1874, employs a diction curiously similar to Nerval's (and to my own) in describing our relation to the speech of others. We may have good reason to cling to our private language, if Jackson ([1874] 1958, vol. 2, p. 132) is right in claiming that we experience others' language as an uncontrollable intrusion into our world: "To receive a proposition and to form one are plainly two very different things . . . when we *receive* a proposition the process is entirely automatic and . . . we cannot help receiving it." As Julian Jaynes (2000, p. 97) puts it, "To hear is actually a kind of obedience." But I continue with the Jackson quotation: "When anyone says to me, 'Gold is yellow,' I am, so to speak, his victim, and the words he utters rouse similar ones in me . . . the revival occurs in spite of me" (p. 132). For Jackson, we are not merely the recipients, we are the *victims* of others' language. In Jackson's terms, the language of others is what might be called an aggravated form of perception, since, for him, perception itself is already an intrusion into our privacy.

In the effort to rescue some part of language for the self, Hopkins (1959, p. 125) formulates the notion that a part of every word is "a word to oneself, an inchoate word," which is at the same time "the idea itself." Jackson ([1874] 1958, vol. 2, p. 201) arrives at a similar formulation by distinguishing between "a subjective word" and "an objective word." Following Kaja Silverman, Jeremy Tambling (2000, p. 25) points out that even voice itself has two tracks: "that which is anterior to clear meaning and nearly linked to the body, and also what is socialized, sublimated uttering language."

There is a sense, then, in which public language contaminates whatever it touches. Hans Apfelbach (1922, pp. 12–13) speaks of dreams as the privileged province of feeling-thoughts (Denkgefühle); they function most effectively in dream because they are not interfered with (gestört) by word-symbols. Snyder and Mitchell (1999) have shown that, in some respects, language may actually be an impediment to thought: deactivating the language centers of the brain facilitates and enhances some important mental functions (in addition to Snyder's thinking cap [see Macrae, 2008], see Minkel, 2006). In notes of my own concerning dreaming, I find the following: "It is as if the visual element contained its own implicit meanings, but undiminished by and unmixed with actual verbalizations that might violate the implicit meanings of the image"; it is as if dream did not want to distort or confuse its procedures with a fully articulated script. It keeps language in its place, as the secondary function that it is.

But language does, after all, occur in dream; in what sense, then, can we say that dream is spared the intrusion of language?

d. Language Dissociation in Dream

As I have been suggesting, dream turns the tables on language. Instead of being squeezed through the toothpaste tube of language, experience in dream simply uses language as part of its repertoire whenever and in whatever way it chooses: in fragments, "correctly," "incorrectly," or not at all. When we sleep, language is at the mercy of dream. For this reason, language in dream can be effortless, whereas, if we try to channel all our experience through language, as we do when we are awake, language, though required for expression, also becomes an obstacle to expression. It always sets up friction in the mind; it demands labor of us to produce the right words.

Language in dream is not required to be responsible; it does not have to follow the rules. As I have written previously (1987a, p. 133), in dream, "one can sometimes feel the bit drop from the mouth of language. In dream, language can be laid down with as little effort as the pavement of a moonlit sky. In fact, far from being produced with effort, language in dreams is a sediment that actually lightens, or relieves the mind-fluid from which it has silted out." In this context, I quote passages in which the satisfaction of dream language is expressed, whether sensual satisfaction, as in "The Song of Songs," or intellectual, as in Ashbery's "No Way of Knowing":

> And the roof of thy mouth like the best wine for the beloved, causing the lips of those that are asleep to speak. ("The Song of Songs," quoted in Massey, 1987a, p. 159)

> Waking up / In the middle of a dream with one's mouth full / Of unknown words (Ashbery, quoted in Massey, 1987a, p. 23),

words that, for all their apparent incomprehensibility, encased as they are in what I have called "the curiously inviolate envelope of dream language" (p. 23), include all the experiences of life.

We fall asleep by getting our language right. To continue the passage I just quoted, from my own text,

> we fall asleep by allowing our words to move with freedom among and around our images, instead of being subordinated to them as they are in our waking life: the words to the accompaniment of which we fall asleep are gradually joined by a dance of images in the background that for once do not (to our relief) exact congruence between themselves and the

language passing through our minds, but let us think what we want. (p. 23)

On the previous page, I write,

> in dreams there are disconnections between images and words; the shuttle does not incessantly flit back and forth between them. In a dream, the image reel can go one way and the sound track another, each pursuing its own course. Often we hear words that do not come from anyone represented in the visual content of the dream, and the words may be completely unrelated to the visual materials. (p. 22)

In film criticism, the term "acousmatic" seems to have been coined to describe those situations in which there is a "separation of the soundtrack from the visual images" (Tambling, 2000, pp. 24, 27).

In *Hourglass*, Danilo Kiš (1990, pp. 145–147) gives us some idea of what it is like to be inhabited by a dream-self while one is going about one's ordinary business:

> sometimes, just for an instant, I catch him in these forbidden, secret activities . . . doing *something different*, something utterly incomprehensible to me, for it is absolutely unrelated . . . to my ideas and thought processes . . . [W]hen . . . I think I've . . . caught him in the act, he slips away into some part of my own brain. . . . He . . . utters incomprehensible syllables, incomprehensible words and sentences . . . [I]f he says them out loud, [he] immediately adds a lot of other sentences or incoherent syllables that make me forget the perfectly clear and unmistakable meaning of the first sentences . . . [T]his Siamese twin of mine . . . actually *thinks* with my brain . . . but not quite clearly or articulately, because one interferes with the other.

At times, such intrusions from the dreaming mind, or possibly even from beyond that level, can occur when one is fully conscious. Once the psychologist Alfred Maury suddenly heard the phrase "analytical geometry in three dimensions" in circumstances having nothing to do with mathematics (Oswald, 1962, p. 107). André Breton (1962, e.g., pp. 34–35) reported similar intrusions. (Incidentally, the 1924 *Manifeste du surréalisme* [in Breton, 1962, pp. 13–63] is still a valuable source of observations and ideas about dreams.) David Foulkes (1985, pp. 38–39) gives the example of the word "Communism" as being superimposed on the imagined action of making pudding in a pan. Arthur M. Arkin (1981, p. 109) reports the case of a subject who heard one word in his dream but spoke a different word aloud; it was as if two people were trying to say the same thing, each in his own way, since the emotional content of the two words was the same: one was "Burglars!," the other, "Mama!" Some patients suffering from jargonaphasia—the use of neologisms and inappropriate words—come up with expressions that straddle the border between the irrelevant and the startlingly relevant: one man said that he had been doing a lot of *verbal gambling*; a woman patient (who had a boyfriend) complained that she had been suffering from a *fressary of my mouthpiece* (Weinstein, 1981, p. 141).

Now, by no means can it be taken for granted that language in dream can ever be entirely detached from the rest of the dream content. At times, dream language seems to have no relevance whatever, but it is hard to prove a negative. What is clear is that it sometimes *seems* to be disconnected from the other aspects of the dream, in the sense that it shows no obvious relation to the action and the images, or, at the very least, that, while dreaming, we do not understand its relevance to the dream, even though we may be able to demonstrate that relevance afterward. There is a relaxation of connection, sometimes to the point of total incoherence, so that we may hear neologisms or nonsense expressions in our dreams that are not experienced as inappropriate or in any way inadequate (see Subsection f, below). Arkin (1981) speaks of "cognitive subsystems that may emit speech independent of the 'executive ego'" (p. 290), and of Broca's area "running on in isolation" (p. 362; see also Pinker, 1994, p. 311, and J. W. Brown, 1981, p. 1). In dreams, we may babble or jabber at will; and, as I have said above, much of the satisfaction or ease of sleep lies in that freedom. It is perfectly all right to be dreaming about an apple while the word in your mind is "orange."

An observation of Dr. Mark Solms' may cast some light on this situation, at least for a professional neurologist. In conversation, Dr. Solms remarked that the dorso-lateral prefrontal cortex (DLPC) is inactive during REM sleep, when much of our dreaming occurs. (Perhaps this fact could help to account for the explicit nature of some sexual dreams.) The deactivation of this "executive" area presumably implies that the control of the relationship between language and images in sleep is also relaxed. Whether that relationship is then taken over by a different set of circuits that imposes a different set of rules, or whether the deactivation of this cortical area is entirely irrelevant, I am in no position to decide. It has been shown, though, that the DLPC is involved in "tasks involving the manipulation of semantic representations" (Martin and Freedman, 2001, p. 243), so that the reduction of its activity should have some bearing on language. Still, since dissociation of language from event in dream is only sporadic, and since the cessation of DLPC activity during REM seems to be continuous, this explanation of bizarre language sequences in dream is at best incomplete. Besides, it would not work for dreams that occur during NREM sleep. Arkin (1981, p. 58) mentions that it was not during REM but usually during sleep stage 3–4 that he encountered "A type of sleep-speech bearing striking resemblance to severe aphasia in the wakeful speech of the brain damaged."

Morris Moscovitch (1973, pp. 114–122; cited in Arkin, 1981, p. 298) offers another intriguing suggestion concerning the source of divergences such as those that I have been describing. He theorizes that, while we are conscious, the left hemisphere of the brain suppresses or inhibits the right hemisphere's speech impulses, but loses control during sleep. Supporting evidence for this assertion is the tendency of the nondominant hand to move during sleep (Arkin,

1981, p. 299). J. Allan Hobson (1988, p. 268) and Frank Heynick (1993, pp. 220–221) speak of the possibility that the right hemisphere may become dominant during sleep, while both Jouvet (1999, pp. xii, 67, 73–74) and Panksepp (1998, p. 142) remark that barriers between different systems are lowered at that time. In fact, Moscovitch's conclusions are in keeping with John Hughlings Jackson's ([1874] 1958, vol. 2, p. 192) much earlier hypothesis that "destruction of the function of a higher centre is a removal of inhibition over a lower centre," so leading to "the increased dischargeability of the right half," as manifested by nonpropositional, emotional vocalizations. Moscovitch's ideas are also in keeping with the competition— the "neural Darwinism" model—that later became popular (see, for instance, Deacon, 1997, p. 457; for a review of the ambiguous evidence concerning lateralization during dream, see Hunt, 1989, pp. 168–174, as well as Antrobus and Bertini, 1992, pp. 15–140).

We all want to be able to fall asleep. Emmanuel Levinas ([1948] 1985, p. 30) goes so far as to call sleep not the antithesis, but the very defining condition, of consciousness: "La conscience est le pouvoir de dormir. Cette fuite dans le plein est comme le paradoxe même de la conscience." ["Consciousness is the capacity for sleep. This escape into completeness is the very paradox of consciousness itself."] At times, we have no choice but to fall asleep. When Emilia asks Desdemona, after Othello calls her a whore, "How do you, my good lady?," Desdemona answers, "Faith, half-asleep" (*Othello*, 4.2.96–97): not indifference, not narcolepsy, but the only possible adaptive response to an overwhelming assault. There are some things that mere consciousness is not equipped to deal with. By a buried ligament of irony, this scene is bound to a later one; Desdemona's desire for sleep is fulfilled by that symbol of rest, the pillow that stifles her.

Actually, as Kraepelin (1906), Arkin (1981), and Jason W. Brown (1979) all remark, and as I have said above, the language of dream bears some resemblance to the language of neologistic jargonaphasia, in which lesions in particular areas cause people to begin to use words inappropriately, or to put together sounds that sometimes

70   do, but sometimes do not, resemble actual words (Kertesz, 1981, particularly pp. 66–68, 110). If it is a language that they are using, it has to be a private language. Clearly, sleep does impose a form of aphasia, in the sense that, during sleep, our speech apparatus is inactivated, and even when we do speak aloud in our sleep, what we say shares some features with the speech of aphasics (J. W. Brown, 1979, p. 38, and Arkin, 1981, p. 358). To help account for such anomalies, it would be important to study the behavior of the language centers during sleep, especially at sleep onset and during REM, by means of brain imaging techniques. I have been unable to find a study of this sort; the technical difficulties to be surmounted are, of course, substantial. Robert Stickgold of Harvard and Allen R. Braun of the National Institutes of Health have expressed to me interest in conducting such a study, but so far without sequel, and Pierre Maquet has told me that, as of May 2008, he knows of no fMRI studies of the language areas at sleep onset.

e. Language in Schizophrenia and Aphasia

So far, I have been dealing primarily with the topics of aphasia and dream speech, but the unusual speech patterns of schizophrenics have also been an object of close study. In the tradition of R. D. Laing, Janusz Wróbel, in *Language and Schizophrenia* (1990), attributes a superior form of logic and realism to schizophrenics' speech. A more recent work, Richard Bentall's *Madness Explained: Psychosis and Human Nature* (2003, pp. 386–400), also describes some striking characteristics of schizophrenic language. Since I am concerned with the relation between the nonstandard speech of sleep and/or of the neurologically impaired, on the one hand, and of poetry, on the other, it would also seem appropriate for me to devote some attention to schizophrenic expression.

Unfortunately, although one might assume that schizophrenic language is different in obvious ways from aphasia, blind studies of the differences have yielded poor results. Two typical papers communicating such negative results are those of Arvin F. Oke et al.

(1993) and Raymond Faber et al. (1983). However, one might say that both the aphasic and the schizophrenic do sometimes strike at a deeper truth. I have already cited one or two instances of what one might call, in a loose sense, metaphoric aptness in aphasic language, which Wróbel (1990, e.g., p. 13) often finds in schizophrenic language. I will not attempt to deal with schizophrenic speech separately, primarily because the topic is too general. Schizophrenic language is heterogeneous, running the gamut from the apparently ordinary to the totally obscure, with stretches of cogency or hyper-expressivity in between. Consequently, it is very difficult to isolate a syndrome or, alternatively, a recurrent pattern of superior expression within this linguistic class for comparison with poetic language. A mental condition such as Tourette's syndrome, with its well-defined features, lends itself much more readily to such a comparison, and, as I have mentioned, Ronald Schleifer has addressed himself with great thoroughness to that subject; I will return to it below.

f. Dream Language and Poetic Language

There are two ways in which language can achieve relaxation. One is the way that I describe at length above: by entering into dream, where language is effortless and need not be responsible or dominant. The other is by an earned relaxation, in the perfection of poetry, when things come together, as someone—perhaps Housman—says, with the finality of a box snapping shut. Yet both outcomes share a similar background; both presuppose what one might broadly call "free association": the dream, almost by definition, and the poem, too, for the perfect rhyme or the perfect line is notoriously unattainable by deliberate effort, but requires at some stage a letting go, a relinquishing of conscious control.

Still, poetry is not an escape from communication, but a pursuit of its ideal form. This is just as true, if not more true, of so-called "nonsense" poetry, which teases and delights us with its apparent irrelevance, only to show that even the most outlandish words or sounds can, like the unicorn, be cajoled into taking up residence in the field of public meaning. The real question, of course, is what happens in that gray zone where language is at once haphazard and meaningful, seeming to face two ways at once (as some might say of the works of Velimir Khlebnikov, of Kurt Schwitters, or of Dario Fo). Even Wordsworth was concerned with this dilemma, though, in his case, the choice was not so much between sense and nonsense as between what was important only for himself and what had universal resonance. The problem is not identical with the one that I am trying to define, but it is clearly related to it. Wordsworth ([1800] 1950, p. 20) says, "I am sensible that my associations must have sometimes been particular instead of general . . . [and] that my language may frequently have suffered from those arbitrary connections of feelings and ideas with particular words and phrases, from which no man can altogether protect himself." Yet Wordsworth thinks it best not to tamper with his language on the advice of critics, since the final arbiter of that crucial distinction between the private and the universal has to be the poet himself, "for his own feelings are his stay and support" (p. 21).

g. Tourettism and Poetry

People who suffer from mental or emotional disturbances may experience this conflict between private and public elements in their speech in an aggravated form. What Wordsworth fears most is that his private associations will deprive his words of the universal significance that he considers essential to poetry. Yet, as is commonly recognized, and as I have said before, there is an aspect of poetry that draws on the least conscious and most private levels of expression to produce its effects. Therefore, it seems reasonable to assume that the relatively uncontrolled speech of Touretters, of fluent aphasics, or of certain schizophrenics, may throw some light on the poetic process. In "The Poetics of Tourette Syndrome: Language, Neurobiology, and Poetry," the seminal article I mention at the beginning of this chapter, Ronald Schleifer (2001) pursues some of the possibilities revealed by this insight. Schleifer does not deal with actual mad writers (such as the eighteenth-century poet Christopher Smart); he is more concerned with showing that poetry in general shares certain features with Tourettism, such as the compulsive or ejaculatory expressions that we think of as typical of Touretters' speech. Schleifer emphasizes what he considers the physical aspects of poetry, where language "means" not by the traditional content of words, but by effects that he believes move us directly: rhyme, repetition, "echoing," and puns (p. 571), what used to be called "primary process" language: the way words rub up against each other, in an almost bodily jostling. (Perhaps one should not ignore the fact, though, that the devices which Schleifer identifies, when they are not supporting the right words, become mere frigid gestures and produce no emotional response whatsoever.) The distinction between linguistic and motor activity is presumably reduced in poetry, producing an "articulate but not quite apprehensible meaning" (p. 577), as intentional behavior is superimposed upon unintentional behavior in poetic language (p. 573).

Schleifer's account, which I find compelling despite some reservations, contains much that is plausible, but it is not difficult to play devil's advocate with his argument. (The following observations are not to be taken as detracting in the least from my admiration for Schleifer's work.) I must confess that when it comes to analyzing a particular poem, such as T. S. Eliot's "Ash Wednesday," Schleifer fails to convince me that there is any resemblance between what he calls the "primal cries" of monkeys (p. 571) and Eliot's words; just try "The Word without a word, the Word within / The world and for the world" (quoted in Schleifer, p. 576) on a monkey, even on one of David Smith's wonderfully introspective monkeys (see, e.g., J. D. Smith et al., 1997)! Nor do I see in Eliot's poem the faintest trace of the incongruous automaticity that is presumably basic to Tourette language. Rhythmic repetition and other "physical" effects there are in abundance, but they are never out of control; they serve only to reinforce the meaning of the poem's words. If anything, these effects work to create the *illusion* of irrationality.

Of course, any poet is constantly exploiting and manipulating the natural music of the language in order to intensify its meaning. Perhaps some poems, such as lullabies, do produce a trancelike state through a kind of physical rocking motion akin to a musical rhythm or chant; this hardly qualifies as Tourettism. As for "nonsense" poetry, as I intimate above, it is an especially controlled kind of poetry: if it were in fact nonsense, no one would bother to look at it. The observation that one can never know exactly what a poem means is perhaps accurate, but it has little to do with Tourettism. The fact that one may not know exactly what Touretters mean does not make their speech poetry.

At one point (p. 571) Schleifer makes (p. 571) reference to a style of writing, revived in the 1990s, known as "Language Poetry"; this is, in turn, linked to "sound poetry," in which the sounds of language become the raw material of poetry, and to aleatory electronic verse. (Much of what has passed as new in this field is third generation surrealism, though not to be taken less seriously for that reason.) At this point, I would like to draw a distinction between Schleifer's undertaking and my own. Schleifer is interested in demonstrating the similarities between a particular linguistic disturbance and the style of poetry. Characteristic, for him, would be the abrupt, ejaculatory, explosive quality of a line from T. S. Eliot that he quotes: "the garden in the desert / Of drouth, spitting from the mouth the withered apple-seed" (quoted in Schleifer, p. 577). I myself am more interested in the drift toward inwardness that I find in all language, and that haunts poetry in particular. Again, Hopkins' (1959, p. 125) meaning for "the self": a kind of laziness, in which language threatens to slip back into dream. For me, a characteristic line, coming close to this border, would be Shakespeare's "Curtsied when you have and kissed / The wild waves whist" (*Tempest* 1.2.378–379), with its multiple alliterations; in its dreamy, relaxed movement, the casual, almost nonsensical perfection of the line, which has succeeded in putting even the sea to sleep, recalls the ease of hypnagogic language. I would even be prepared to cite, in this connection, the occasional language poet Susan Howe (2002): "A portable altar strapped on his back / pure and severe / In the forests of Germany he will feed / On aromatic grass," where an inherent incongruity in the language is made to feel plausible, though one is left searching for a principle that holds it together: grappling with the "Private Enigma in the Opened Text" (Davies, 1984, p. 7). In poetry, the right words give one access to the nonverbal.

Perhaps, then, there is an undercurrent of aphasic or inchoate dream language in all linguistic expression, not just in language poetry: a drift that ends at the horizon of jargonaphasia.

h. Dreams and Metaphors

I return to the subject of language in dream, not, this time, in relation to poetry as such, but in relation to the general problem of metaphor. To what extent can dream language be said to be metaphoric? I will proceed gradually.

All discussion of dreams labors under a handicap. A sort of Heisenberg effect sets in when we start to talk about dreams: the very act of examining them deforms or transforms them, so that we end up studying our own artifact rather than what was originally there. Even the language that we use to talk about dreams infiltrates and replaces the dream's substance, like formaldehyde or, perhaps, like acrylic (as in London's 2002 "Body Worlds" exhibit, subsequently repeated elsewhere). Dreams are not pathological or deficient experiences. "We murder to dissect" is particularly true of dreams, those most fragile and vulnerable of our psychic efflorescences. This effect is particularly noticeable when we isolate an aspect of dream, such as language. Only the dream as a whole makes sense, even if it is not the kind of sense that we usually think of as "sense."

Recently, under the tutelage of my daughter, I was attempting to learn how to use a disk in an unfamiliar computer. The following morning, in a half-dream, a computer disk appeared, but this disk contained, in their total intricacy, the suffering and melancholy beauty of all human relationships. Together with that image, the phrase "Human suffering and condensation" appeared in my mind.

Once isolated, the phrase "Human suffering and condensation" looks laughably like an entry for calling up information on a computer. The global quality of the dream experience, in which the words were part of an extremely moving dream, was completely destroyed: removed from its native milieu, the phrase is merely awkward and stupid. The computer disk, too, considered from a waking point of view, was also a poor vehicle for the burden of meaning that it carried, yet for the purpose of the dream it was completely adequate.

Such an observation draws one inevitably into a consideration of metaphor in dream, although the subject is tangential to my interests in this chapter. Metaphor usually passes through language; for that reason alone, as well as for many other reasons, it has its place in a discussion of language in dream. Nevertheless, I think that the argument that dreams always work out a linguistic/propositional precursor is quite wrong. Dreams are expressions of fear or desire, not regurgitations of statements. The fact that we can resolve a dream into a propositional structure does not mean that we constructed it that way.

It can hardly be denied that dreams produce something that resembles metaphor, but the differences are also striking. The most obvious difference is that dream metaphors (as in the above example of the computer disk) do not have to be good; if they happen to be "good," it is not because the dream was trying to make them good by following some aesthetic criterion. Dream metaphors have no audience, and they are not meant to be judged by a reader. A second difference is that dream metaphors do not invite us to speculate endlessly on their implications. We think of a good metaphor as radiating multiple meanings; its beauty is reflected in its teasing out our interpretive impulse. It is an invitation to endless rumination. Dream metaphors are monosemantic; in this respect they are the opposite of metaphors. They do not vibrate with the tension between tenor and vehicle. Even if we do not know what they are saying, they are in principle univocal and unequivocal.

Dream symbols (if that is what they are) are not caught up in the dualities of consciousness. The interpretation of a dream is meant to stop dead when the interpretive act is completed; it does not present the spectacle of an endless negotiation between the conscious (the tenor) and the gift or contribution of the unconscious (the vehicle). It is not that the distance between tenor and vehicle is collapsed in dreams (cf. States, 1988, p. 134); it was never there in the first place. This is why most dream images, when considered as metaphors, seem unsatisfactory: they were not metaphors to begin with. Dreams may display the vividness of metaphor, but without the metaphoric pattern of stresses, they remain occult images.

To return to my dream of the computer disk, as I have said, the "vehicle" in this case seems to be a poor one for the conscious observer. But, in dream, not only is the vehicle not trying to be adequate to the tenor, but in principle it can never be adequate to the tenor. It may be adequate, though, to something greater: to the extralinguistic thought and feeling that it embodies. An example from Danilo Kiš (1990, pp. 148–149) will clarify my meaning:

> All that was left in my mind was the impression of a nightmare, all that I could formulate coherently was a single word: BIG, an adjective coupled with some impossible thing, with some concept that I could not identify but that emanated unconscionable horror. Yet . . . this word was totally adequate, it fitted in quite naturally and logically with it in gender, number, and case, even though this concept was still outside the confines of the intelligible.

The dream word or image, truncated though it may seem, conveys the entire experiential context of the issue with which it is trying to cope. A specific mood or thought/feeling (cf. Apfelbach, 1922) is represented by a wide array of images, words, or events that have only an indirect relation to the original mood or thought. Yet, in its totality, "A dream extracts what is essential from the disordered chaos of life," though "[o]nce the dream is over, its richness, its logic, its clarity become the nothing after an explosion" (Herling, 1967, p. 52).

i. Thought and Expression

This section may be considered a philosophical digression, but some may find it of interest for exploring the status of language in dream. As I have said once before (see Subsection b, above), if dreaming, in some form, preceded language, then there is no reason why it should engage in syllogistic thought, or why it should make its contents available to consciousness in discursive form. In fact, if some part of dream is still prelinguistic, then dream may exhibit the struggle to make the transition from feeling and unformed thought (sometimes called "mentalese" [Pinker, 1994, p. 78]) to formed expression. This may seem obvious, but the process by which that transition is carried out is anything but obvious. It may be that the initial leap from the formless to the formed can take place only by a special device: perhaps a joke, or even a pun. As I have written before (1976, pp. 84–85),

> The pun may be the only form in which thought can emerge directly into language. Thought must leave itself behind before it can become sentences. If there be a way in which thought can get into language untranslated, it would have to be through some metaphoric process. The pun seems to me at least to exemplify the kind of mechanism by which thought might become words directly . . . [I]t avoids the necessity of working a thought over into the serial or temporal procession of a grammatical statement. . . . Perhaps the pun also symbolizes the direct translation of thought into language because it depends on a physical device (sound) to produce a change of context, and thought can become words directly (rather than by moving through some intermediate process that reduces it to temporal forms, logic and grammar) only

by a trick that projects the globe of abstraction onto the
plane of reality. Before thought becomes sentences, it must
become things.

(My examples were drawn from Lewis Carroll's *Alice* books.)

When thought gropes its way into articulated experience, it
has no guide: there is nothing to tell it what forms to assume. On
striking the barrier that isolates it from expression, it bursts into a
spray of possibilities or images, something like an energetic particle
striking its target in a cyclotron. The readings that can be obtained
from the impact do not offer a direct representation of the parent
particle, but they enable us to infer something about its properties.
The words, images, and actions tell us something about the thought
that gave rise to them, but they are not a direct representation of
that thought, since there can be no such thing as a direct representation of a thought. As Vygotsky (1970, p. 150) says, "thought does not
have its automatic counterpart in words." The leap into language
is necessarily awkward, the leap into an imagistic medium only a
trifle less so. If what is produced at that point in a dream can be
called a metaphor, it is a metaphor without a specifiable tenor.

Writing about this dilemma in her chapter on Kant, Angelika
Rauch (2000, p. 110) reminds us that, for Kant, "an idea has no direct, sensuous representation," and that it is essential to distinguish
between mental representation [*Vorstellung*] and material representation [*Darstellung*]. The difficulty is that there is always more in the
mind than there is in words: "Semper mens est potentior quam sint
verba" (Blumenberg, 1974, p. 24). A metaphoric process is involved
in "the transfer or transference between intuitions and affective
language" (Rauch, 2000, p. 110). But since emotional forces "cannot
be immediately contemplated, much less grasped in any cognitive
form, there are only the imaginative representations" (p. 78). "[W]
e actually do not know . . . the affects in themselves; we only know
them through the accompanying, forever newly created images
with which the mind, involuntarily, tries to represent sensuous
experience" (p. 79). Rauch tries to account for the choice of these

images in a general way, as the mind's blending of Freudian associations with the sediments of what she calls tradition. I confess that I myself do not know how the mind arrives at the particular images that it gives off in the attempt to accommodate thought/ feeling to expression.

A dream is a parade of partial embodiments of a mood, each trying to solve the same problem set, each approaching it from a different angle, and inexhaustible in its available supply of images and actions. Again, as Angelika Rauch puts it, "the goal . . . is to achieve a mental representation, that is, to impose (cognitive, conceptualizable) form on the amorphous mass of sense impressions and affects (*Sinnenreiz*)" (p. 82). The psychoanalyst imposes a vocabulary on the turmoil of the emotions; the dream puts that state forward in images/actions that, as best they can, capture and give shape to what is happening. The dream is also an imposition, but it is more respectful of the possibility that precision is not the highest value in understanding (see my *Find You the Virtue*, 1987a, p. 172). The dream does not falsify in order to express; it does not fall back on discursive language, but it does simplify the wordless chaos of thought and feeling. It gives those elements a local habitation, if not always a recognizable name. The dream does not attempt to supply titles for the inner experiences that have given rise to the dream, but confines itself to images associated with those experiences. I quote Rauch again:

> Because feeling does not have a form, it has to be treated like an inner sensation which can be understood in terms of the images it triggers. These images do not, however, represent the feeling as such, for they are independently existing representations or fantasies that are merely associated . . . [I]t is not the mental representation of a rose that is pleasant but the images remembered *along with* the subject's affection by the rose. (pp. 87–88)

When we fall asleep, feelings replace things as the only available objects of our perception. Since consciousness continues during sleep, and since it has been left with nothing but these inner emotions and sensations to work on, it scans the landscape of our feelings and gives them perceptual embodiment as best it can; it tries to visualize the invisible.

When we are awake, a full confrontation with the complexity of our feelings is perpetually deferred by the linear nature of language. Like a plunger or piston, it shoulders experience aside as it bores through our lives. But, as Edward Bradford Titchener points out, even in ordinary communication, meaning often has to be distributed over verbal and nonverbal media: "meaning is carried by all sorts of sensational and imaginal processes" (quoted in Heynick, 1993, pp. 166–167). Language in dream seeks to combine all of these. It attempts the impossible and conveys the impossibility of the attempt. Consciousness itself is sometimes identified with language; it is therefore bewildering to watch dream language as it tries to function within the regime of the unconscious—as the junior partner, once more, in an ancient firm. It is a heroic but complicated effort at adaptation, and it will be a long time before we have identified all its features, clarified its purpose, and defined its achievement.

j. Meaning and Aphasia

But to consider again the comparison between aphasia and dream language, for which Danilo Kiš' hopeless struggle to get beyond the word "BIG" in his dream might serve as a good example, what imposes limits on the analogy between language in dream and the aphasias is the fact that, in aphasia, language, no matter how incomplete, is still the dominant vehicle for thought and expression, whereas, as I have said above, in dream, language is subordinated. Dream bodies forth experience; language channels and confines it (cf. Blechner, 2001, pp. 25, 28). Dream is a global event, in which language has only a limited role. As I have said repeatedly, dreams are textural in meaning, somewhat like experiences themselves; they create a diffuse feeling, in which particular words and im-

82   ages may appear, but it is only in their entirety that they produce the dream atmosphere.

With these reservations in mind, I will return, though with great hesitation, to the topic of the possible affinities between dream language and aphasia. To be completely scrupulous, one must acknowledge that, in most situations, one cannot be sure whether the neologisms produced in the aphasic condition are merely the wrong sounds attached to the right words, whether they are meaningless gibberish, or whether they express some private meaning that cannot be conveyed in any other way. "Will no one tell me what she sings?," asks Wordsworth ([1807] 1965, line 17), almost in desperation, about the Solitary Reaper's song. I will not go into Gregor's twittering in Kafka's "Die Verwandlung" ["The Transformation"] ([1915] 1970).

In connection with the question whether the language of jargonaphasia has meaning for its speaker (see Subsection d, above), I will try to muster a few examples from my own experience, fragmentary though they be. In one case, I dreamt a clause that began, "knocked off my feet by." The context suggested completion by a phrase something like "waves of words"; instead, I dreamt the concluding phrase "verbal newspapers," which seemed incongruous to me even in the dreaming state. In another case, I dreamt that some nations had reverted to policies that favored smoking. The content of my thought at that point was, clearly and specifically, "When nations go crazy . . .," but I was aware in my dream that the words "when nations get together" were the words that were actually accompanying the thought; I was also aware that they were inappropriate. I knew what I was thinking, but I had somehow been "given" the wrong words to express the idea. Another time I dreamt that I was planning to catch large fish, but the words accompanying this expectation had me "writing numerous plays." On still another occasion, during a slight indisposition, I dreamt the words "Kill a Japanese any time and." These words expressed the first part of a perfectly normal "if-then" construction, such as "If you carry an umbrella, you won't get wet," or "Drink too much and you'll have an accident." Unfortunately, I cannot recall the "if-then" idea that my sanguinary language was intended to express; all I could remember was that it was mild and commonplace, and that it had nothing whatsoever to do with either the Japanese or murder. The only feature of the dream thought that was retained by the dream language was the grammatical, syllogistic construction. But in this case the words, inappropriate as they were, did seem adequate to express the perfectly normal thought that I was having. They clung to it at least as closely as normal language would have done; they too were wedded to the idea. As in the case of the example from Danilo Kiš that I quote above, they said exactly what I was thinking, even though they bore no apparent relation to it. As Kraepelin remarks repeatedly of his own dreams (1906, pp. 71–77), there was no struggle to find "better" words. It was as if the *overtones* of the words harmonized to express the idea, perhaps more accurately than the "appropriate" words could have done. As far as I was concerned, the question was settled: for the dreamer/aphasic, nonsense words can mean, and express fully, the thought to which they are connected. I had subjective proof of the hypothesis that "the jargonaphasic knows what he is talking about" (Peuser and Temp, 1981, p. 290).

To return to the context of Wordsworth and Kafka: if the purpose of the postcollective arts is to encourage me to say what I, and only I, can contribute to human experience, then a private language says it best, even if, paradoxically, no one will be able to tell what it means. That is the language of dreams.

One theory of jargonaphasia (e.g., Kertesz, 1981, p. 107) holds that certain brain lesions inactivate the auditory system that monitors how what we are saying sounds to us. What happens then is that we cease to know that what we are saying is wrong. This condition, known as anosognosia, or ignorance of one's own disease, is a basic feature of jargonaphasia, but it has been argued that sufferers sometimes show an underlying unease with their condition that betrays a partial awareness of the defect (Weinstein, 1981, p. 145; Solms and Turnbull, 2002, pp. 266–271).

84   In dreams, though, nothing can be wrong: we are, for the most part, completely at ease with our own language. Dream language may not be poetry, not even "language poetry," but it is a use of words that can provide us with complete satisfaction. In dreams, we achieve a coherence that consciousness cannot hope to imitate. The point at which the machinery of the mind seems to have broken down completely is also the point at which the unity of the self has been accomplished. At a horizon beyond "Jabberwocky," beyond any poetry of sound, lies the ideal place where our "anosognosia" is total; where we have no difficulty in accepting our own language; where the sounds we utter, whether privately or aloud, are coin of the realm. There we are ourselves, as we were to begin with.

2. LITERARY USES OF NEUROSCIENCE

Having addressed a very specific issue in the first part of this chapter—the relations among aphasia, dream, and poetry—I will now deal with some of the more general features of the field spanning literature and neuroscience. The first is the use of literature as what one might call a stable of examples illustrating neurological processes. The second is the practice of using neurological materials as the subject of literary works.

An egregious instance of the first practice may be found in Paul M. Matthews and Jeffrey McQuain's *The Bard on The Brain: Understanding the Mind through the Art of Shakespeare and the Science of Brain Imaging* (2003), in which Shakespeare's plays become exhibits in a psychology textbook. Matthews and McQuain gloss Macbeth's attempt to clutch the imaginary dagger as "an illustration of what each of us does scores of times each day: reaching and grasping," a "critically important activity [that] occurs in the cerebellum" (p. 73). The authors go on to write, "Thus, when moving his right hand, an actor playing Macbeth would activate the right cerebellar hemisphere and the left primary cortex, as shown in figure 14" (p. 75). Matthews and McQuain seem strangely unaware that you do not need Shakespeare to make this point, or that using a Shakespeare play to illustrate this process does nothing for either Shakespeare or neurology.

Even a thoroughly sophisticated author such as Mary Thomas Crane, in *Shakespeare's Brain: Reading with Cognitive Theory* (2001), finds it necessary to make a passing gesture toward this kind of thinking. She tells us that

> the formation of a sentence—"Whether 'tis nobler in the mind to suffer / The slings and arrows of outrageous fortune, / Or to take arms against a sea of troubles, / And by opposing, end them,"—for example, probably involved activity first in the occipital, posterior superior parietal, and posterior inferior temporal lobes, central to the generation of mental images, and then in the perisylvian cortex (regions of the brain located near the sylvian fissure, also called the lateral sulcus), where the images (slings and arrows, arms, sea) and concepts (grounded, perhaps, in a Lakoffian metaphoric structure, "life is a war") would be associated with appropriate words and formed into a grammatically acceptable sentence. (p. 15)

I fail to see why Crane finds it necessary to inform us of these facts, or what they could possibly contribute to our understanding of Shakespeare. (For a considerably more interesting discussion of Shakespeare in a neuroscientific context, see Davis, 2007.)

Another approach to the neuro-psychological study of literature can be found in Susan Nalbantian's *Memory in Literature: From Rousseau to Neuroscience* (2003). Nalbantian tries to discriminate among the various forms of both memory and forgetting described by authors over the last two centuries, and she aligns each of these versions of the memory-forgetting complex with a different neurological theory of memory. Of course, our grasp of the memory process is still uncertain; Douwe Draaisma, in *Metaphors of Memory: A History of Ideas about the Mind* (2000), is one of those who has emphasized the inadequacy of current theories. As Draaisma (2004,

p. 12) says of psychology in general, we tend "to give preferential treatment to questions suited to the instruments at our disposal." Nevertheless, certain features of mnemonic activity are well recognized and distinguished from one another, such as Baddeley's "sketch pad" for short-term memory versus long-term storage, even though the underlying physiological processes may still be in dispute. (For a review of developments in the neurochemistry of memory, see Kandel, 2006. Still more recently we have the development of concepts such as "clustered plasticity.")

Again, there seems to be a fairly clear-cut distinction between spontaneous memory (as in Marcel Proust's recalling his childhood on tasting a biscuit) and purposeful recall. Such sharp distinctions are not always applicable, though, and, in the end, Nalbantian follows the strategy of matching the aspect of memory that dominates in a given work with the theory that best explains it. So, for instance, Edmund Rolls' hippocampal engram theory (p. 139) might work well for Proust (p. 141), whereas a biochemical approach, such as that proposed by Eric Kandel, Bryan Tully, or James L. McGaugh (p. 143), would do better for Baudelaire (pp. 142–143). For the Surrealists, Jean-Pierre Changeux's theory of childhood encoding might be apt (p. 172). I myself am moved by the thought that the unforgettable scene at the end of Tolstoy's *The Cossacks* ([1862] 1965), where the Chechens are cornered in an empty clearing—a "sacred space"—and about to die, is so powerful partly because a piece of the surround, a piece of that sacred space, attaches to the reader's memory (Edward Rolls' theory, in Nalbantian, 2004, p. 140); or, as we can see some things best on the periphery of our vision, maybe the scene is so powerful because memory is displaced to the surround. (The strange Donald Thompson case, widely reported in the 1990s, is an example. See Greenberg, 2008, p. 10.) Perhaps, though, it is simply that Tolstoy is a great writer, and no one else could produce the same effect merely by using the same psychological ingredients.

In any case, in the present fragmented state of memory research, we seem to be in the position of the blind men trying to determine what an elephant really is: a rope (the trunk), a column (the leg), a fan (the ear), and so forth. Rubin and Greenberg (2003), as well as John Bickle (2003), have found the role of memory in personal narrative at least as hard to understand as it is in fiction. Certainly, memory in literature (like memory outside literature) is subject to division and classification. Beyond that, though, as Borges (1962, p. 59) says, we may be "coining the faceless wind." Memory in literature has always been the refuge of subjectivity. It seems only fitting that it should continue to elude our grasp.

As has happened repeatedly during the course of my investigation, though, no sooner have I written a sentence of the sort that says neuroscience has no access to a particular problem than I find that someone has indeed attempted to grasp the ungraspable, to pin Proteus down. Andrew Elfenbein's "Cognitive Science and the History of Reading" (2006, especially pp. 487–489), for instance, exemplifies the cognitive sciences' efforts to track and capture the elusive memory function in literature. There are interesting suggestions in Elfenbein's analysis, but predictable difficulties also arise. Like Elfenbein, one may observe how normal memory processes function in reading; however, almost by definition, literature, being an art, is not normal. We "process" it in uncontrollable ways, choosing to emphasize what we prefer. But, even allowing for the possibility that certain elements of the reading process follow universal psychological laws, not everyone may be willing to engage in the particular analytic strategies required by Elfenbein's, or any other, cognitive-psychological approach. First, one would have to accept the presuppositions of cognitive psychology and the "models" (in Elfenbein's case, the "landscape model") that it generates. Second, one would have to commit oneself immediately to learning its terminology ("cohort activation," "coherence-based retrieval," "reinstatement inferences," etc.). It is something like reading late Blake, or Freud: love me, love my system—a full commitment is required. There is no room for the eclectic critic here.

It may be argued, of course, that any scientific psychology, or any psychology that aspires to the status of a science, will require the acquisition of the relevant vocabulary. The question, as I have said above, is whether one is willing to accept the presuppositions of cognitive science as one's own point of departure, and whether this approach yields insights sufficiently important and sufficiently different from what can be understood without it to warrant the effort. Here the individual critic will decide. (A more user-friendly example of the ways in which cognitive science can be applied to the problem of memory in literature is the article by Hirsch and Hirsch [2006].)

The fact remains that memory is a deeply embedded element of the literary process, and one can barely talk about literature without addressing problems of memory in some way. One of the difficulties is that literacy itself can distort the literary experience, especially if one goes about deliberately "studying" a text. The possibility of going over a written or printed piece of writing creates an artificial "memory"; it enables one to discover coherences or to decide that certain details are especially important, whereas a single hearing or even a single reading may produce no such results. We produce artifacts in the place of the original work by destroying our capacity to forget, to pass lightly over what was not meant to be salient. Everything acquires the capacity to become equally important, and we create organic unity where the opposite may have been intended. For instance, I have always thought that emphasizing plot while teaching Dickens is a mistake. I remember how my parents read Dickens: for the sake of the episodes, above all for the sake of the characters; the plot was a distant afterthought. One read Dickens for the sheer delight of what was going on from one scene to another; what held the reader was the crazy tone, not the laboriously created frame. To produce a seamless web in the place of chaotic, wonderful preposterousness is to replace Dickens by total substitution.

Another feature of memory in literature might be addressed by psychology or, perhaps, even by neurology, but I do not have the means to approach it in any but aesthetic terms. It has to do with the function of forgetting within memory itself, with the way in which forgetting is embedded in memory. Perhaps recollection requires that everything except what one is recalling at a given moment be suppressed; otherwise, everything would crowd upon everything else, and one could not know what one was thinking at that moment.

The role of forgetting in memory was discussed with respect to children's stories at the April 2006 Conference on Literature and Cognitive Science in Storrs, Connecticut. The question was asked: Why do children who know the outcome of a story with a dramatic or, possibly, a frightening ending show increasing rather than decreasing terror with each retelling of the story—especially when they themselves have demanded that the story be retold?

I myself have been thinking about the role of forgetting in relation to poetry that one has memorized. One does not appreciate a poem less when one has learned it by heart; in fact, it is presumably because one likes and values it that one has taken the trouble to commit the poem to memory in the first place. What is it that we like about poetry? It is the shock of realization that comes when it dawns on one that that next word was exactly the right word, that it did something unexpected but perfect. On a first reading, we may not achieve this awareness: there is too much to absorb; there are too many judgments to make. So we memorize or, perhaps, gradually absorb the poem in order to savor it at our leisure. Once we learn it by heart, though, in principle its words should no longer take us by surprise, or catch us unawares: the very fact that we can repeat them from memory proves that we already know the next word before we have spoken it. Yet there is that same hesitation, that same moment of unawareness before the next word, that same fleeting blindness, as though we had forgotten the word just before we articulated it, which guarantees its novelty. We "know" the poem in outline while we are repeating it, but we actually know the poem word by word, only as it comes back to us word by word. We are not really in control: the poem retains its independence even after we "have" it completely. It is the poem that chooses or refuses to return to us, not we who bring it back.

I always wonder at the unquenchable freshness of the lines

"Thy light alone—like mist o'er mountains driven, / Or music by the night-wind sent / Through strings of some still instrument" (Shelley's "Hymn to Intellectual Beauty," [1817] 1956, lines 32–34); for some reason, they always crop up first when I think of passages that I have learned by heart. Perhaps these lines provide a particularly good illustration of my point because they contain so many words pronounced as one syllable—sixteen, I believe, out of a total of twenty-one. It is as though Shelley doles them out to us as we repeat them, almost teasing us; there are more pauses than one would expect in the lines, spaces like those between dots of colored candy on a sheet. The impression is driven home by the even slower, more widely spaced distribution of the monosyllables in the final line of the stanza, the "punch" line: "Gives grace and truth to life's unquiet dream" (line 36).

How one might represent such a process neurologically, I have no idea, though I believe that Professor Moshe Bar's brain imaging lab at Harvard is studying the role of forgetting as a necessary ingredient in memory. There are, of course, other subtle events taking place in these lines that are equally difficult to account for. The actual meaning of individual words, or even of entire phrases, is elided as we read as well as when we remember; one slips over words (e.g., "night wind") without really knowing what they are while in the very act of affirming their rightness. Again, there is a double activity going on: if it is not forgetting while remembering, it is not knowing while knowing, and both are necessary.

~~~

My second subtopic in this section, the use of neurological problems as a subject in literature, is less complex. The recent popularity of neuroscience has led to its becoming a common theme in literature, but mental and emotional difficulties, whether understood as originating in physiology or circumstance, have always been a staple of fiction and poetry. A good example is one of Robert Louis Stevenson's major works, *Strange Case of Dr. Jekyll and Mr. Hyde*, which may have been influenced by Pierre Janet's theory of multiple personalities (cf. *The Three Faces of Eve* and Ian Hacking's *Rewriting the Soul: Multiple Personality and the Sciences of Memory*, [1995]). Stevenson himself denied any such influence, but a paper by Dr. Anne Stiles at the 2006 Robert Louis Stevenson Conference in Saranac Lake seems to confirm it beyond a reasonable doubt. I quote Richard Dury's summary in the minutes of the conference:

> Anne Stiles . . . placed *Jekyll and Hyde* in the context of contemporary theories of the brain and personality. The "paper . . . in a French scientific journal on sub-consciousness" that Fanny [Stevenson] says partly inspired the tale may well have been on the Félida case reported in a series of articles in the *Revue scientifique* in the 1870's, but Stiles suggested that Stevenson may have learned about this and other famous cases of dual personality through articles in *Cornhill Magazine* in 1875 and 1877 (in the same period in which Stevenson's *Cornhill* articles were appearing, from 1874 onwards), where double personality is related to theories of the double brain. Unbalance of the two hemispheres was seen as leading to mental illness including dual personality disorder and the right hemisphere was associated with a series of negatively connoted personality-traits (the instinctive, feminine, savage, criminal and insane!), all of which can be seen in Hyde. The cases discussed in the articles (Félida, Sergeant F.) have interesting affinities with Stevenson's "Strange Case": the painful transition from one consciousness to the other; the morally less-controlled second state, and Félida's change from morose to cheerful (like the first transformation to Hyde). Stevenson even uses some of the language of the scientific article but ironizes the detachment of the scientist by making the doctor and patient the same person.

In our own time, the dilemma created by commissurotomy (the severing of the link between the brain's hemispheres) was picked up by the philosopher Roland Puccetti and incorporated in a fictional work called *The Trial of John and Henry Norton* (1973). The famous Polish science fiction writer Stanislaw Lem deals with a similar problem in *Peace on Earth* (1994). Schizophrenia is, of course, the subject of the movie *A Beautiful Mind*, though madness as a topic in literature and art goes back at least as far as the Greeks. (I need not mention that other movie that plays with neurology, *The Eternal Sunshine of the Spotless Mind*.) The "idiot savant" is most familiar through the film *Rain Man*. I am told that in Mark Salzman's *Lying Awake* (2000), a nun with a temporal lobe tumour has religious seizures but is also driven to write poetry. The protagonist of Mark Haddon's novel, *The Curious Incident of the Dog in the Night-Time* (2003), displays the literalism of Asperger's syndrome. Laura Otis has translated the stories of the famous neurologist Ramón y Cajal (2001), and in *Membranes: Metaphors of Invasion in Nineteenth-Century Literature, Science, and Politics* (1999) and *Networking: Communicating with Bodies and Machines in the Nineteenth Century* (2001), she traces the metaphors of the neural net and the communications network through nineteenth-century novels from *Middlemarch* to *Dracula*.

Robert Cohen's *Inspired Sleep* (2001), if I understand its complex plot, seems to deal with a pharmaceutical company's testing of neurotropic products on an unsuspecting public. One of the characters in Jonathan Franzen's *The Corrections* (2001) suffers from Parkinson's disease. The protagonist of Ian McEwan's *Saturday* (2006) is a neurosurgeon. But perhaps the most striking of the books that I have seen that has a neurological issue at its center is Jonathan Lethem's *Motherless Brooklyn* (1999). The protagonist suffers from Tourette's syndrome, and much of his language is meant to reflect the peculiarities of Tourette speech. There has been so much interest in this condition that a theater festival, called NeuroFest, devoted to dramatizations of Tourette's syndrome and related ailments, was organized in January 2006 (McElroy, 2006).

A secondary form that has developed alongside what might be called the neurological novel is the neurological case history treated as a literary genre. Most readers would think of this genre as having originated with Oliver Sacks, but it actually goes back much farther, to Alfred de Vigny, in *Stello: A Session with Doctor Noir* ([1832] 1963), to Karl Albert Scherner, to Emil Kraepelin, and to Freud. Antonio Damasio, Harold Klawans, Mark J. Blechner, and Paul Broks are other distinguished practitioners of this style (see also Oates, 2007, p. 48).

3. METAPHOR

I will now try to deal with the perennially perplexing topic of metaphor. Although antagonism to metaphor is an ancient phenomenon in ethics, theology, and philosophy, in the later twentieth century a systematically antimetaphoric school of thought developed in literature, again exemplifying what I refer to above as the iconoclastic tendency in twentieth-century thought. This attitude is clearly expressed in the essays of the novelist Alain Robbe-Grillet (1963). Not long afterward, in his famous essay "La mythologie blanche" ["White Mythology"] (1971), Jacques Derrida made a determined effort to dislodge metaphor from its pride of place in initiating thought.

For me, the last of these assaults is associated with the most disturbing possibilities from a neurological point of view. What if we could, in fact, get rid of metaphor? Despite Derrida's powerful objections, I remain convinced that metaphors are incubators of ideas. Metaphor acts as a springboard for an idea, or contains the foreshadowing of an idea. We know that with Asperger's syndrome, or with damage to the right frontal lobe, the capacity to apprehend metaphors, or to think in metaphoric terms, is impaired. I believe that the implications of this observation have not yet been confronted squarely.

The value of a metaphor depends largely on the unpredictable associations that it creates, associations that reverberate in unforeseen ways. If we can, by mechanical means, cut out the metaphoric function, we have cut out thought: perhaps not quantifying thought, but certainly the creative thought, whether artistic or scientific, that keeps us all alive. Another way of putting the issue is to ask: If we wanted to restore the metaphor-generating function to a brain from which it had been excised, what, exactly, would we be trying to put back? Could we put the potable tear back in the hand of Elizabeth Bishop's "Man-Moth" ([1946] 1977)? Could we help him find his way back through the subway? If one were trying to restore someone's lost capacity to name carpentry tools, or even the lost capacity to employ function words, one would know exactly what outcome to look for: but what would a capsule of imagination look like? What would be the thing that we were reinstating? The unnamable source of naming, the indescribable source of description? For, if we cannot say what imagination is, or from what follicles the ova of metaphor are born, how can we know what it is that we are trying to restore? The essence of humanity, or some cubic centimeter of protoplasm in the inferior parietal lobule, as Ramachandran et al. (2005; see also Graham, 2005) would have it?

I leave this problem for more competent philosophers to deal with. To return to the matter of metaphor in literature: I would like to point to some features of metaphor that relate it to neurological processes that *have* been closely studied. One of these is the so-called "Appeal of the Rare" (Sapolsky and Ehrlich, 2003). As Neil Greenberg (2004, p. 8) has said, "The organism . . . possesses a fundamental need to identify novelty and its possible salience." The retinal cells themselves, without even having recourse to the brain, grow inured to their environment and lose focus when exposed to consistent patterns, but respond strongly to the unexpected (Unger, 2005; for more on unconscious vision, see Goodale and Milner, 2004). In the words of Robert Sapolsky and Paul Ehrlich (2003, p. 62), "A selective responsiveness to rarity and contrast seems to be a feature of our sensory systems, a phenomenon called contrast enhancement. . . . Suddenly throw in something different and the sensory system comes alive." The advantages of having a sensory system that is kept "alive" are obvious. A good metaphor is a way

of *creating* "contrast enhancement"; it jogs our somnolent senses
into alertness with the turn of a word. At least since Coleridge, in
the early nineteenth century, it has been recognized that the more
unlike the two terms of a metaphor are, the more effective that metaphor will be: not "See the fingers of the birch twigs in the spring,"
but "See the chocolate mist around the birch twigs in the spring!"
(Coleridge, 1957, vol. 1, entry 1783).

The second connection that I wish to make between metaphor
and a neurological or, rather, neurochemical process is considerably
more daring, or, perhaps, merely far-fetched. I have been wondering whether the metaphoric function could have any relation to the
behavioral influence of oxytocin.

The literature on the effects of oxytocin is as voluminous as it
is contradictory. One researcher in the field, Rebecca Turner, in a
February 9, 2004, email to me, declared flatly that "there is almost
no data on oxytocin in humans in terms of its effects on behavior,
cognition or emotion"; she adds that the inability of oxytocin to
cross the blood-brain barrier makes experimentation extremely difficult. Evidence from long-standing experiments by Swiss researchers, who have reported quite extraordinary results in enhancing
trust among subjects who have been given oxytocin to sniff, seems
to me to contradict Turner's opinion (see, e.g., Kosfeld et al., 2005,
and Wade, 2005). One thing, however, seems to be clear: oxytocin is
associated with sexual functions, from orgasm to parturition to lac-
tation. Pleasure in nursing seems to be related to oxytocin; Johnson
(2003) mentions that "oxytocin's effects are heightened by estrogen"
(p. 76) and that there is evidence "that oxytocin is embedded in our
reward circuitry" (p. 75). At least in animal models, it also seems
to reduce separation anxiety and to promote bonding (Panksepp,
1998, pp. 272–273, and Moberg, 2003, p. 66 and passim).

Even more important for my purpose is the suggestion that oxytocin helps one to respond favorably to new situations and frees one
to learn. A curious observation by Shelley Taylor may have some
bearing on this possibility: Taylor (2003, p. 75) interprets the high
levels of oxytocin found in women with unhappy marriages as a

sign that they are ready to look for more suitable partners. If oxytocin does indeed facilitate an acceptance of unfamiliar situations, as Walter Freeman (2000) and Moberg (2003, p. 66) maintain, it could enhance the "appeal of the rare" that, as I have pointed out, is inherent in metaphor. More than that: I have always thought that metaphor is associated with some aspect of the erotic. A moment of excitement and pleasure accompanies the creation or discovery of a successful metaphor; there is a marriage of unforeseen, one might say exogamous, elements in every metaphor. Furthermore, when one is actually in love, one is in what might be called a state of continuous metaphoric perception: nothing is merely what it is; everything is immediately transformed into something more significant.

I will quote what I have written in another context (1987a, pp. 8–9) concerning metaphor and the knowledge that, as a subspecies of love, depends on metaphor:

> [T]he erotic state is privileged. Love sees double, depression single. When one is in love, a thing strikes the senses *initially* as that which it is not (arm rather than branch, jingling coins rather than rattling leaves, etc.). Connotative perception precedes denotative perception. Perhaps the tendency to perceive in metaphors is a secondary sexual characteristic of *homo sapiens*, and may even have survival value for the species. . . . Knowing is a subspecies of metaphor. . . . As Wordsworth says, "where he [man] has no pleasure he has no knowledge" (preface to the *Lyrical Ballads*). It may be said that perception itself depends on love. "In order to see anything—a leaf or a blade of grass—you have, I think, to know the keenness of love" (John Cheever, "The Fourth Alarm").

That throwaway remark by Cheever's protagonist is buttressed by Nicholas Humphrey's (2006, pp. 47–72) very specific demonstration that perception becomes available to us only through the affective link of sensation, which (as in "blindsight") is demonstrably different from perception as such. Jaak Panksepp (1998, p. 315) emphasizes the dependence of perception on emotion. Of course, the general distinction between thought and feeling has always been difficult to maintain. In an essay that he entitles "Emotional Abstractions," L. Dugas (1905, p. 485) states that "feeling leads and directs thought; it provides the connectivity of thought; what we call a generalization is often nothing more than a series of images unrolling around a feeling." ["Le sentiment conduit donc et dirige la pensée, en forme la trame; ce qu'on appelle une idée générale n'est souvent qu'une succession d'images se déroulant autour d'un sentiment."]

In the erotic state, one simply *sees* more. The constant novelty created by the flow of metaphoric perception associated with sexual love yields a powerful benefit, a strongly enhanced awareness of and openness to novelty. The neuroscientist Walter Freeman associates these conditions with oxytocin. William Benzon (2001, p. 81) quotes Freeman's claim that, especially at moments of social reorganization, oxytocin dissolves "preexisting learning by loosening the synaptic connections in which prior knowledge is held." To quote Freeman (2000, p. 418) directly, oxytocin "appears to act by dissolving preexisting learning. . . . This opens an opportunity for learning new knowledge" and is capable of producing the "born-again" experience. Whatever the mechanism by which oxytocin achieves this effect, Moberg (2003, pp. 72–73) also finds that oxytocin facilitates learning.

If even some of these observations on the behavioral effects of oxytocin are right, then the possibility that metaphoric experience is associated with an oxytocin "high" begins to seem plausible, whether or not an experimental confirmation of the connection could ever be made. Metaphor is beneficial because it combines an enhanced alertness to contrast or "the rare" with (if my hypothesis has any validity at all) an increased availability for new knowledge or experience that is presumably advanced by a neural facilitator such as oxytocin. At the same time, I must confess that my choice of oxytocin for that role was dictated by the lack of an ideal candidate. In a recent conversation, Dr. Bruce McEwen of Rockefeller

University said to me that, although some neuroscientists claim such effects for oxytocin, and possibly for prolactin, in his opinion no substance clearly and unequivocally dedicated to promoting the acceptance of novelty has been identified. (Perhaps the point is that oxytocin reduces anxieties associated with novelty rather than promotes its acceptance.)

As for alcohol and antidepressants, they may reduce inhibition and/or stress, but they do not particularly help us to absorb the new. Obviously, a positive response to novelty is a complex function that may recruit more than one circuit and require more than one chemical; dopamine is often mentioned as an element in the process. In *Hardwired Behavior: What Neuroscience Reveals about Morality*, Laurence R. Tancredi (2005, p. 133) remarks that the nucleus accumbens is particularly susceptible to activation by the anticipation of unpredictable rewards, which leads to the release of dopamine, which in turn produces "excitement, euphoria, and the desire to take additional risks." I have found suggestions (e.g., Prolo and Licinio, 2002) that at least one gene, the dopamine receptor gene DRD4, may influence how much people seek out new experiences; however, a 2002 study (Kluger et al.) casts doubt on the association. According to an article by Frédéric Limosin et al. (2003), the DRD1 gene appears to be connected with novelty-seeking, at least in alcohol-dependent men. (This article also includes a brief review of other studies that examine genetic factors which encourage or inhibit the pursuit of novelty.) Lawrence Wilkinson of Cambridge has provided another genetic clue: "the mother's Nesp55 gene . . . may encourage the mice to take more risks on behalf of the group, whether that risk involves looking for food or defending the group" (Zimmer, 2006, p. D6). Disabling the gene "makes mice less likely to explore a new environment" (Zimmer, 2006, p. D6).

Be that as it may, to return to my much earlier point, it is apparent, at the very least, that the knowledge function and the metaphoric function are closely intertwined. This fact raises major difficulties for the neurological approach to the mind. How can we ever "get behind" thought by neurological measures, if we cannot seize its taproot? As long as the element of metaphor, the element of the unpredictable, the ever-elusive "creative" process, are acknowledged to be essential for thought, as long as thought cannot be confined within strictly mechanical procedures, neurology may be said to have only a limited role in understanding the mind.

> I shake off the sun,
> and tramp down
> to where it washes up
> as soft and fine as swift
> as fingers at the keyboard
> of musical my toes. (Grieco, 2007)

In this poem about the sea, what possessed my friend Peter Grieco to dare to violate the language in that outrageously ungrammatical, totally unforeseen, perfectly tuned last line?

Yet the analysts of our subtlest mental processes also tramp on, undeterred. When we recall that Allen Snyder and Mitchell (1999) have shown how closing down the language functions can improve our ability to draw (a discovery that was, I think, foreshadowed in Massey, 1987a, pp. 115–118); that Alice Flaherty (2004) believes she has located the neural source of writing mania; that Tim Griffiths (Zimmer, 2005), among others, has begun to localize the source of obsessive musical hallucinations (see also Sacks, 1987, pp. 138–143; Nagourney, 2001; and Zimmer, 2005, on the work of Dr. Victor Aziz), we must begin to wonder, after all, how far neurology may try to go.

But, even if much of what happens in our minds can be correlated with physical processes, it must also be remembered that most of what is in our minds is not simply *in* our minds, but exists between us and among us. It would not make much sense to say that *King Lear* was simply something *in* Shakespeare (cf. R. A. Wilson, 2004). In one sense, consciousness is private; in another sense, it has to be part of the world: the world that comes into us, the world of the communication that goes on between us, and the world of exchange that we call perception. For the present, at least, the answers to the questions, of what do our inspirations consist? from where do they arrive? to whom are they delivered? how do they relate to others? how can we describe them? and, above all, what do they mean to us? remain, mercifully, beyond our grasp.

But knowledge is, after all, just part of our evolving construction of the universe. For Aristotle (see McKeon, 1941, p. ix), for Kierkegaard (*Concept of Irony*, Part II, [1841] 1968), and for Heidegger ([1927] 1967, p. 38), the future always means more than the present; the potential always stands higher than the actual. Einstein puts it more simply: "Imagination is more important than knowledge." In the end, our responsibility is not to learn to know what we imagine, but, as Shelley said, to learn to imagine what we know.

4

Music and Language in Dream

1. INTRODUCTION

The new field of neuroaesthetics (within which the above topic is lodged) has been expanding rapidly during the last few years, and neuroscientific conferences and lectures have begun to proliferate. An annual conference on neuroaesthetics is now held at Berkeley, under the auspices of the Minerva Foundation. In September 2005, there was a conference on cognitive neuroscience and the arts in New York, sponsored by the Dactyl Foundation. There is a regular public lecture series at Harvard on the same subject. A large meeting on Literature and Cognitive Science, with a neuroscientific component, was held in April 2006 at the University of Connecticut. A major conference on music and the brain, with published proceedings, is held more or less annually at various international venues; abstracts from the proceedings of the 2006 Bologna conference are available at http://www.icmpc2006.org/. Among relevant conferences in 2008 were one on the neurology of music, held in London and organized by F. C. Rose, and one on music and language, at Tufts University. Moreover, an Academy for Neuroscience and Architecture has been established at San Diego, and there are institutes on neuroscience and music at McGill University, at the Université de Montréal, and at Harvard.

Jonah Lehrer's *Proust was a Neuroscientist* (2007) sheds light on some interesting phenomena in the neurology of music. As I mention in the preface, Lehrer proposes that the audience's rioting at the first performance of Stravinsky's "Rite of Spring" may be accounted for by the response of certain neurons to unfamiliar sounds. According to Lehrer (pp. 141–142), when these neurons fail to find a pattern in the sounds to which they are being subjected, an excessive discharge of dopamine can follow, producing something like schizophrenia. Whether or not Lehrer's explanation holds up under scrutiny, and one can think of several caveats, it implies that a neurochemical approach to an aesthetic problem is possible.

In view of the strong interest in the neural substrate or accompaniment of music, among the other arts, it seems appropriate to revisit a musical phenomenon that has until now been described only anecdotally, in order to see what light may be shed on it from a neurologically informed perspective. This phenomenon is the musical dream, long familiar in musical circles, but regarded as vanishingly rare by the general public. These dreams are less unusual, however, than the public thinks (Prokop, 1979; Streich, 1980; Grace, 2002; and in a personal communication with William Domhoff, December 7, 2004). They occur more often in people with musical training, but they are not confined to musicians (Prokop, 1979, and Streich, 1980).

One can encounter an expression of interest in the musical dream in the psychological journals as early as 1964 (see Knobloch et al., 1964), but it was after the major article published in 1978 by the Stanford psychologist Roger N. Shepard that a steady trickle of publications concerning the musical dream began to appear. Of course, the phenomenon had been known long before to anthropologists and musicians. As associated with rites of passage in some early societies (see Herzog, 1936, and Nettl, 1986), as a theme in fiction (see Tolstoy, [1865–1896] 1991), or as reported by composers (see Wagner, 1978–1980), the musical dream was understood as occupying a significant intermediate ground between psychology and aesthetics.

Besides being of interest as a phenomenon in itself, as a form of "endomusia," the music in dreams is of major theoretical significance because it occupies a unique position in the hierarchy of dream elements. In a word, music is the only faculty that is not altered by the dream environment, whereas action, character, visual events, and language may all be modified or distorted in dream. In addition, the relations between language and music, the two main aural elements of dream, have important implications that remain to be formulated.

In this chapter, I intend to review, in approximately chronological order, some of the literature on the musical dream, and then to consider the subject in its own right, since I believe that there is much to be learned from it; the musical dream is not merely a curiosity, but a potential source of valuable information. I will also review the history of some debates on the primacy of music or language, since they have a direct bearing on the question of the differences in the way music and language behave during dream.

I will divide these reviews into seven subtopics: hearing music in dreams is a fairly common experience; music is the only faculty that does not undergo distortion in sleep; music survives sleep more easily than words do; in the word-music combination, music overrides words; music may work to the advantage of the right hemisphere, which probably becomes more active during sleep; the combination of words and music may constitute a distinct module; the fact that the right hemisphere is not committed to discursive language may have something to do with music's dominance over words— again, in sleep as in waking.

2. HEARING MUSIC IN DREAMS IS NOT AN UNCOMMON EXPERIENCE

As has been pointed out by Herzog (1936), in some societies the musical dream is thought of as a universal experience. Herzog (p. 318) found among a group of Indian tribes, the Pima, the Yuman, the Shoshoneans, "and other tribes along the Pacific shore, . . . the song series conceived in supernatural dream-experiences." For the Yuman, "music is almost exclusively an individualistic activity; *all* singing is the result of the individual's supernatural dream experiences" (p. 320). The Pima, on the other hand, do have some songs "which cannot be remade, or rather redreamt, by individuals, but must be acquired through teaching" (p. 320). For some cultures, then, in order to sing, one must first have dreamt one's song.

Roger N. Shepard (1978) reports some of his auditory dreams in considerable detail. One night he fell asleep while thinking of Bach's "Toccata in F Major" and "suddenly heard the pipe organ, in all the richness and majesty of reality, thundering forth with such force that I awoke with a start" (p. 178). On another occasion, he dreamt of a baritone saxophone, and "Each click of key, each peculiarly breathy, husky tone and deep 'squawk-like' transient unique to the saxophone registered itself with crystal clarity. Then . . . a second sax joined the first in an almost unbearably exquisite syncopated counterpoint" (p. 178). Again, he dreams a harmonica and clarinet duet, or another organ toccata (p. 178). Shepard observes that much of what he hears in dreams is music that he has never known before (pp. 178–179). He then goes on to give the partial score of two of his musical dreams, which he reconstructed to the best of his ability after waking up (Figure 10). Shepard's account of his numerous musical dreams is unusually specific in its recording of instrumentation and timbre.

The next article I want to mention, by Heinz Prokop, appeared in 1979 under the title "A Creative Product of the Unconscious" ["Eine schöpferische Produktion des Unbewussten"]. It begins with the usual references to the dreams of composers such as Mozart and Handel, but it also points out that Cosima Wagner's diaries (1978–1980) contained very specific references to Richard Wagner's musical dreams. Since Wagner seems to have been in the habit of telling Cosima his dreams in the morning, there is a great deal of material to sift through. Glancing at the diary entries just for the year 1878, I myself found nine mentions of musical dreams. On Tuesday, April 2 (vol. 2, p. 56) there is even mention of the specific notes that Wagner heard during one of his dreams. On Tuesday, December 24 (vol. 2, p. 237), he had two musical dreams. Cosima herself dreamt that she heard her first husband playing piano (Feb-

Supplement from Shepard (1978)

A. Reconstructed beginning of fugue heard in a dream

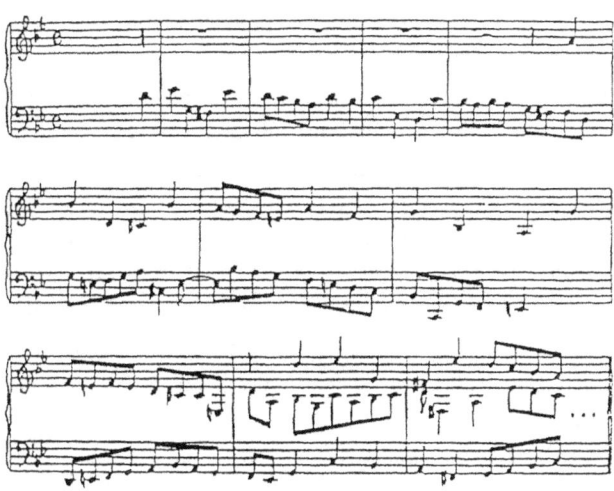

B. Reconstructed piano music heard in a dream

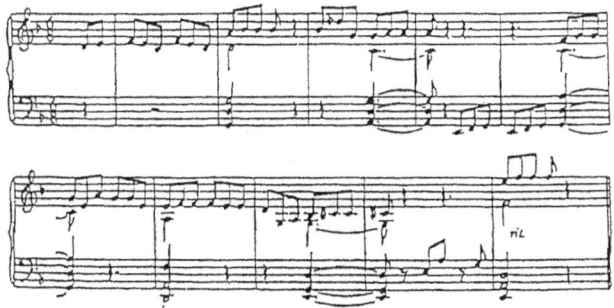

FIGURE 10. Partial score of two of Shepard's musical dreams. In Shepard (1978, pp. 179–180).

ruary 3, 1877; vol. 1, p. 945). Elsewhere she makes the astute observation that any account of a dream must be approximate: "More or less thus, for no dream can be told quite as it was" (vol. 2, p. 184).

Prokop (1979, p. 49) identifies twenty-four of Wagner's dreams that contained unequivocal musical experiences and sixteen more concerning musical performances. Among the twenty-four unequivocal musical experiences, he identifies "four-hand sonatas, plain-chant, an Italian choir, church music, symphonic music by Mozart, piano music by Chopin and twice piano music played by Hans von Bülow, Cosima Wagner's divorced first husband" (p. 50). Interestingly enough, the extracts from Wagner's own compositions appear in his dreams in a playful context.

Prokop also includes Bruckner's report of a dream in which he heard a theme of his seventh symphony: "This theme wasn't mine at all. One day the conductor Dorn, an old friend of mine from Linz, appeared to me in a dream and dictated the theme to me: I wrote it down right away. Pay attention, Tonerl, Dorn added: this will bring you success" (p. 50). Prokop also reports musical dreams of both Robert and Klara Schumann, and that Brahms dreamt the successful revision of his failed youthful symphony, which became the piano concerto in d minor (p. 50). Prokop also reports at length a December 25, 1815, musical dream of Rahel Varnhagen's (p. 50)—Varnhagen was one of the famous salon women of the early nineteenth century—and quotes several major German authors who reported musical dreams. Eduard Mörike dreamt a concert with unusual instruments. Friedrich Hebbel recorded a dream in which his early beloved Elise played a golden harp so splendidly that she was "herself enraptured" (p. 51).

According to Prokop, "When one happens to ask friends or patients whether they have musical dreams, one often finds that such dreams occur much more frequently than one thinks" (p. 51). He reports in detail a friend's dream of performing a theme from the Grieg A minor sonata for piano and cello, and he also describes his own frequent musical dreams, which involved various musical instruments as well as vocal music; a male choir; on one occasion, "a wonderful deep black basso, accompanied by a women's choir singing high above"; and, at times, orchestral music in the style of Mozart (p. 51). Prokop goes on to say that "the musical element was almost always far stronger than the visual element of the dream" (p. 51). He also mentions the possibility of studying the musical dream by brain imaging. As I myself will point out below, when one awakes, music is not forgotten as quickly as is the other content of dreams. Prokop continues: "dreams associated with music sometimes remain in the memory for a whole lifetime," or the music "may remain in the memory for some hours before it falls victim to the usual process of dream-extinction" (p. 55). Another observation of Prokop's that I think is universally true is that music in dreams is never confused or merely discordant: "Since in my numerous musical dreams only harmony and not a single case of disharmony has appeared, it would interest me to know whether dreams with discordant music can appear at all" (p. 55). He cites humor in musical dreams, but never disharmony.

Prokop tells us that during a fairly long period of his life he himself had musical dreams quite regularly. He would hear distant music when falling asleep and when awakening.

~~~~

A third article, by Hildegard Streich, appeared in 1980. In the introduction, Streich suggests that "[t]he musical contents of dreams are often related to the archaic significance of music as a cathartic harmonizing power" (p. 12). She speaks of music as a life force, using the example of a Danish composer who dreamt of himself lying in a coffin as a corpse, then hearing music, and finally coming back to life (p. 13). One of Streich's patients, who was severely inhibited, dreamt of "natural, heartfelt" black music protecting her from "persecution" by whites: "there was a long-drawn-out sound that made the earth shake" (p. 14). In the "Music and Love" section, Streich mentions a patient who lacked feelings and heard "in dream a wonderful melody, and with it the text, 'Music takes the shortest time to love again,'" which the patient experienced as a message of hope that he might overcome his lethargy (p. 15).

Streich describes the experience of a scientist who had the same musical dream repeatedly when he was about twenty: "Two altos sing in parallel thirds a wonderful heavenly music. It is of a rapturous and indescribable beauty" (p. 16). One of her depressed patients dreamt that "[w]e were standing in a festively illuminated concert hall, with music in our hands.... The song celebrated the arrival of spring in its full power.... It recalled the Song of Solomon" (p. 16; the interpretation of this dream is predictable).

Music in dreams is also understood as the domain of wordless understanding: as Wagner puts it, "As pure organ of the Feeling, it [music] speaks out the very thing which Word-speech in itself can not speak out . . . *the Unspeakable*" (quoted in Taylor, 2007, p. 109). A music teacher dreamt of being among students from various countries: "An Indian woman is particularly friendly. I sing a psalm to a very free melody. When I have finished, she says to me in English: that is extraordinary: she has understood every word, even though she doesn't know the language" (p. 17). The same woman dreams of

> [v]ery intense music in a big, half-dark, cathedral-like space.
> An organ, or something like it. Music, as a single great stream
> of sound from my life, including everything that pertains to
> me as well as all the forces that have formed me.... At the
> same time there is still other music, the sound of other organs,
> intermingled with it.... When it comes back, I say "That's it.
> Now one can hear it clearly." (p. 17)

Toward the end of her article, Streich adds, "The examples given here are a small selection from the author's collection of more than 700 dreams with musical content. These are dreams of people from many different parts of the world. From musicians and non-musicians, from musical and so-called unmusical people . . . from the practice of various analysts and from the author's own psychotherapeutic and music-therapeutic practice" (p. 18).

The conductor David Blum (1996) has left us an extremely moving account of his musical dreams during a period of treatment for metastatic cancer, two years before his death. As he puts it, for him his "Musical dreams unrolled like a film" (p. 28). Blum's dreams seemed to provide suggestions as to how to manage his moods or cope with depression: "the dreams were setting their own agenda" (p. 30). But in one dream, he was told to conduct a concerto and refused; what he wanted instead was "an intimate piece of chamber music" (p. 30)—a Mozart quartet. At another time, he dreamt that the Mozart clarinet concerto was being played too quickly, and he realized that he needed to slow down. Entering the radiation "dungeon," as he puts it (p. 30), he was accompanied by the third Leonore overture, suggested to him in a dream. Throughout this process of triumph and agony—as one could indeed say of all musical dreaming—Blum did not feel that he was the agent in what was happening: "I am not the doer, but the receiver" (p. 30).

From my own experience, I can add a few details that I have published previously (1987a, pp. 114–115):

> It is night—I am sleeping in a small hotel on the island of Cyprus. I wake up and write down a melody that has come to me in my dream [the tune in Figure 11]: a procession of young people, dressed in beautifully colored robes, has gone by, singing.
>
> In the morning, I tell my companion the dream. (There is barbed wire separating the Greek and Turkish sectors of Nicosia, but the worst of the war is still to come.)
>
> I am at my summer home in Roman Valley. In my dream I am watching an opera with an acquaintance, who says—or I say—while the music is in progress, "I've no idea what it is."

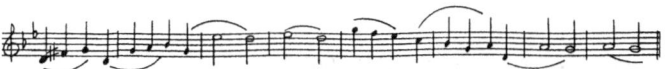

FIGURE 11. *Melody heard in dream by the author.*

A very thin, slight, swarthy singer with a beautiful, melting bass voice begins a simple melody, in which he is joined at the end of the second bar by the choir. . . . I retain only the first few notes of the aria. . . . But frequently, I am so discouraged by the complexity of the music that I am hearing, or am about to hear, in my sleep, that I deliberately stop the dream, in despair that I do not possess sufficient musical knowledge to write down what is coming.

Like the other components of dream, music in dream arrives uninvited and, one might say, uninvented. When one composes consciously, one *listens* to the music that one is about to play or write down; in dream, one only *hears* it. One night, while half asleep, I was trying to get myself to hear dream music, priming the process by thinking up an orchestral counterpoint to the melody that I was attempting to generate. None of this was working very well. But in the very midst of this vain effort I became briefly aware that I was *hearing* another melody quite unlike the one that I was trying to make up: a lively, buoyant melody, in striking contrast to the lugubrious bass line I had been struggling to produce. My other mind was doing something too, but something completely unrelated to the music that I was consciously inventing.

Deirdre Barrett (2001) has written on the musical dream in a chapter titled "The Devil Plays the Violin: Dreams and Music" (pp. 66–81 in *The Committee of Sleep*). Doubt seems to have been cast on two of her examples, the Tartini "Devil's Trill Sonata" and Beethoven's "Tobias" canon (see Stenzl, 1991, pp. 10–12, and Beethoven, 1996, p. 488, fn. 1), but she records numerous other instances of musicians' musical dreams. Among the cases she mentions are those of Paul McCartney (pp. 66–67), Lucy Davis (p. 68), Sir Arthur Sullivan (p. 69), Vincent d'Indy (p. 69), C. E. Hutchinson (p. 70), Bill Barton (pp. 70–71), Billy Joel (p. 71), Joseph Shabalala (pp. 72–73), and Shirish Korde (pp. 75–77). To the list of popular performers I would like to add the fiddler Emile Benoit, who, lacking a recorder, called his sister in the middle of the night and asked her to record over the telephone his violin rendition of a musical dream that he had just had—and that became the piece "Emile's Dream" (M. Miller, 1987). (A related phenomenon is the intrusive musical hallucination, as reported by the Cape Breton fiddler John MacDougall [see Mazurkewich, 2006].) But if "it is very common for people to wake up in the morning with songs running through their heads," as the music psychologist Robert Zatorre reports (quoted in Blakeslee, 1995), then having musical dreams must be a widespread experience indeed. In fact, Nancy Grace (2002) points out that an astonishing number of popular compositions that are assumed to be of ordinary provenance are actually derived from composers' dreams.

Theodor Adorno's reports of his musical dreams, in the recently published *Dream Notes* ([2005] 2007, pp. 48–49), are particularly interesting in light of Adorno's influential writings about music, writings that actually figure in Adorno's own dreams. This short book is based on a small selection of Adorno's dream notes between 1934 and 1969. One of the half-dozen musical dreams is especially striking because it contained music that was doubly imaginary yet nonetheless real. Adorno dreamt that he was playing a tabletop rather than a keyboard, "But there was music. I can clearly remember very powerful, vivid chords in E flat major of the kind that I would like to be able to produce" (p. 54).

I thought that I had reviewed the relevant literature for my subject pretty thoroughly when, in 2006, Steven Brown informed me via email that he had just published an article with a section on the musical dream. Brown (2006, p. 38) makes the by now familiar observation that such dreams reveal "a great deal of creativity lurking within our minds that we can either not access or not externalize during our waking hours." He also quotes the composer Max Bruch's claim that his most beautiful melodies came to him in dreams (p. 38). From his own experience, Brown provides an abundance of interesting detail. He has musical dreams every week or two, consisting of both familiar and original music, but so far he has not succeeded in remembering the latter after waking, though what he does remember (style, tempo, tonality, orchestration, or emotional tone) is certainly substantial. Although he composes only for piano, his dreams are of "richly orchestrated scores" (p. 38) in his preferred classical style. The music consists of continuous passages, not repetitive phrases. The dreams may involve playing an instrument or visualizing a musical score (p. 33). In one case, Brown dreamt an original orchestral piece in the style of Tchaikovsky. Brown wakes every morning to the sound of music in his head, and it takes him a while to realize that what he is hearing is inside himself. These experiences lead him to ask himself "whether there are two distinct music tracks (i.e., waking and sleeping) or simply two modes of operation of the same track" (p. 34). What Brown describes is entirely consistent with the accounts of others who have spoken of their musical dreams.

An article by Uga et al. (2006), which appeared in the same month as my own article on the musical dream (I am grateful to Kurt Ruppel for the reference), provides conclusive evidence for my contention that dreaming music, and in particular original music, is a fairly common experience, especially among musicians. Finally, in a more recent book, Oliver Sacks (2007, pp. 279–284) discusses the musical dream, with particular reference to the dreams of Wagner, Ravel, Stravinsky, and Berlioz. More important, building on his own vast clinical experience and on the work of other researchers such as Jason Warren, Sacks shows that musical hallucinations also occur widely outside dream experience.

The most evocative description that I know of a musical dream is the one Tolstoy, in *War and Peace* ([1865–1869] 1991), attributes to the sixteen-year-old Petya the night before he is killed in battle:

> Ozheg-zheg, Ozheg-zheg ... hissed the sabre against the whetstone, and suddenly Petya heard an harmonious orchestra playing some unknown, sweetly solemn hymn.... The music

became more and more audible. The melody grew and passed from one instrument to another ... [E]ach instrument—now resembling a violin and now a horn, but better and clearer than violin or horn—played its own part, and before it had finished the melody merged with another instrument that began almost the same air, and then with a third and a fourth, and they all blended into one, and again became separate and again blended, now into solemn church music, now into something dazzlingly brilliant and triumphant.

"Oh—why, that was in a dream!" Petya said to himself. ... "It's in my ears. But perhaps it's music of my own. Well, go on, my music! Now!" He tried to conduct that enormous orchestra.

"Now softly, softly die away!" And the sounds obeyed him. "Now fuller, more joyful." ... And from an unknown depth rose increasingly triumphant sounds. "Now voices join in" ordered Petya. And at first from afar he heard men's voices and then women's. The voices grew in harmonious triumphant strength, and Petya listened to their surpassing beauty in awe and joy. (p. 1127; I am grateful to my daughter, Rachel Massey, for reminding me of this passage.)

Tolstoy was, of course, passionately fond of music, and he played the piano competently. It is hard to imagine that Tolstoy would have assigned this dream to Petya if he had not had a similar experience. Nor would it have been the only case where he used a dream of his own as a model for one of his characters' dreams (Gustafson, 1986, pp. 322–324).

3. MUSIC IS THE ONLY FACULTY THAT DOES NOT UNDERGO DISTORTION IN SLEEP

I will now go on to consider some of the theoretical issues regarding the phenomenon of the musical dream. One crucial fact about musical experience in dreams sets it apart from the other elements of dream, that is, from image, action, and language. As I mention in my Introduction, all three of these undergo some form of distortion in dream or, at the very least, exhibit marked irregularities when compared with waking experience. Music alone remains intact in dream (Prokop, 1979, p. 55); indeed, if it were to undergo major deformations, it would not be reported as music, which is, after all, an art form and requires a certain integrity to be recognized as such. One might say that music never sleeps. It is as if it were an autonomous system, indifferent to our consciousness or lack thereof. In my reading, I have encountered only one case in which there is mention of music in sleep being "garbled" (S. Brown, 2006, p. 33), but, in a personal communication, Dr. Steven Brown has subsequently explained that he was using the adjective merely to designate those dreams in which he heard a pastiche of another composer's style.

This is not to say that our other faculties are never capable of normal function during sleep. Deirdre Barrett (2001) makes it abundantly clear that one may engage not merely in competent mentation while asleep, but in astonishingly creative and ingenious mental acts in all fields. (As I have mentioned, pp. 66–81 are specifically concerned with the musical dream.) What distinguishes music in dream from these other achievements is that it is *consistently* normal, whereas normal or superior function in other areas is exceptional or, at least, intermittent.

Allan Snyder has shown that the suspension of certain conscious functions, particularly those associated with language, can release creative potential, and that, during sleep, the mind occasionally takes advantage of this relaxation in order to carry out remarkable acts of reasoning and invention (see Osborne, 2003). What makes music exceptional is that it *always* acts as if the intrusive linguistic element were in abeyance (Massey, 1987a, pp. 117–118), perhaps because music does in fact have deeper roots in human development than does language (Darwin, [1871] 1902; S. Brown, 2000; Molino, 2000; Benzon, 2001; and Huron, 2003).

As we can see from the many cases that Barrett (2001) cites, as well as from my personal experience, music in dream is not fragmented, chaotic, or incoherent (Prokop, 1979, p. 55); neither does it decay as rapidly as do the other components of dream on our awakening. Another fact worth noticing is that, in almost all cases, the music we remember when we awake from a dream does not change in quality in retrospect. It is usually as good or as trivial as it was in the dream, whereas words that seemed striking or beautiful in a dream, and that appeared fully expressive of the feelings they accompanied, are notoriously liable to appear silly or nonsensical if we recall them when we wake up. (This is also true of poetic dreams, despite the much-debated case of Coleridge's "Kubla Khan.")

Music in dreams, then, is the same as music in our waking life. A well-known book on dreams by Frank Heynick (1993) is titled *Language and its Disturbances in Dreams: The Pioneering Work of Freud and Kraepelin Updated.* I do not think it likely that anyone would write a book titled *Music and its Disturbances in Dreams,* since I, at least, am not aware of any such disturbances' having been reported (again, Prokop [1979] bears me out here). Neurological studies by Bernard Lechevalier (1997), Mark Jude Tramo (2001), Tim Griffiths (see "Music on the Brain," 2001), along with those reported by Paul M. Matthews and Jeffrey McQuain (2003, pp. 139–141), show that many different brain functions are involved in musical experience; nevertheless, music behaves as if it were a single package, immune to the centrifugal forces that seize upon our mind in sleep. It is tempting to think that the neural substrate of music may possess some comparable autonomy or integrity.

Dr. Isabelle Peretz of the University of Montreal comes to the same conclusion concerning an independent neural pathway for music. I quote a *New York Times* review of her work:

> Dr. Isabelle Peretz . . . studies patients with lesions in the auditory cortex that impair only their ability to recognize music, while leaving unscathed their power to understand speech, environmental sounds and other acoustic information.
>
> Dr. Peretz's results suggest that the brain has something

specifically designed to process music, although the precise
location or nature of such a . . . keeper remains unknown.
(Angier, 2001)

Dr. Peretz reiterates this view in a more recent article: "recent evidence suggests that music might well be distinct from other cognitive functions, in being subserved by specialized neural networks" (Peretz and Hyde, 2003, p. 362).

Salomon Henschen postulated the existence of a separate center for music perception as early as 1919 (J. W. Brown, 1979, p. 128). Arthur L. Benton (1980) makes similar observations with regard to what are called the "amusias," and, of course, Oliver Sacks (1987, e.g. pp. 18–19, and 2007) emphasizes the apparent independence of the musical faculty from other functions. Eran Zaidel (2001, p. 396) remarks that some split-brain patients who are unable to speak are still able to sing, and the same is true of many stroke victims (Kolb and Whishaw, 2001, p. 341; see also Brust, 2001, pp. 145–146).

4. MUSIC SURVIVES SLEEP MORE EASILY THAN WORDS DO

I have already mentioned another feature of dream music that distinguishes it from the other components of dream: when we wake up, it does not decay as rapidly as do speech, imagery, and action, with the rare exception of the occasional obsessive word or phrase (Kiš, 1990, pp. 148–149; for an attempt at a chemical explanation of the rapidity with which dream content disappears from memory, see Hobson et al., 1998, p. R11, though it is worth noticing that initially forgotten dreams may be remembered much later; Bertini, 1984, pp. 260–261, claims that dream recall is proportionate to left hemisphere activity during REM). As I mention above, it is not an uncommon experience to have the music from one's dream continue running in one's mind after one awakes (Blakeslee, 1995). I assume that the reason why most dream content does not persist is that, if it did, it might cause an animal that could not distinguish between sleep and the waking state to act out dream situations after arousal.

Why would music be different? One possibility is that music simply does not conflict with conscious behavior, as persistent language or imagery might do. Most dream language is notoriously evanescent. Perhaps the difference can also be accounted for by the difference in the form of the two functions. Language, especially in dream, may be fragmented or incoherent, and is often attached to incoherent events in a dream, whereas music is an appetitive, coherent, frequently pleasurable form of action, with a formal contour and inner momentum—a purpose of its own. As I have pointed out above, it is, after all, an art form. Once more, then, it appears that the independence of the musical function may account for the fact that it survives longer than do the other contents of dream. (Evenson [2003, p. A14] quotes Christo Pantev's claim that music is the last thing that we forget, and Oliver Sacks [2007, p. 337] also emphasizes the persistence of musical memory.) One cannot ignore a further possibility, namely, that music may be independent, not only because it is differently localized and differently organized, but also because it is served by electrochemical processes different from those associated with image, language, and narrative, and so may not be subject to the same amnestic forces. (Apparently musical timbre is localized differently than are ordinary speech sounds; even rhythm and meter are neurologically distinguishable [see Belin, Zatorre, and Peretz, as cited by Sacks, 2007, pp. 108–109].)

It may also be worth mentioning that we are apparently not the only species that has musical dreams. The zebra finch rehearses its song during sleep (Dave, Yu, and Margoliash, 1998), rather as rats rerun the mazes that they navigated when awake (Kolb and Whishaw, 2001, p. 474; on bird song, see Rothenberg, 2005). It seems that our (putatively) autonomous musical system has been around for a long time, perhaps, like the siamangs' duets, as a prelinguistic form of social contact (Darwin, [1871] 1902, vol. 2, p. 737; Darwin is emphatic about music's preceding language [see Faienza and Cossu, 2003, p. 393]). If music is a basic function that (given the age

of the earliest musical instruments) possibly precedes even pictorial representation, it would also be the only function that is also
an art form. Even babies are primed to respond to music (Kolb and
Whishaw, 2001, p. 346). Perhaps the musical arts are necessary for
normal brain development in humans (cf. Thaut, 2005, p. 25).

The area of the relation between words and music harbors some
of the most intractable problems in aesthetics. (Fred Lerdahl [2001]
makes an ambitious attempt to define that relation, and Nigel Fabb
tells me that he and Morris Halle are in the course of developing
an alternative to Lerdahl's scheme.) Words have an inherent musical quality; in fact, the rhythm and intonation of a sentence are
probably developed before the specific words are fitted to the tonal
structure. Arkin (1981, p. 343) refers to these as "tone groups" (see
also J. W. Brown, 1979, p. 33, and Anderson, 1981, p. 296). Shelley
actually planned some of his poems in "nee-na" syllables before he
wrote them out (Massey, 1987b, p. 236).

It is good to remember that, no matter how fresh it may seem
to us, this subject, like most others, has gone through a long and
tortuous tradition of inquiry and speculation. "Prima le parole—
dopo la musica! / Prima la musica—dopo le parole!" ["First the
words—then the music! / First the music—then the words!"] were
already slogans, if not battle cries, in the eighteenth century. (For
a recent work on the so-called "Querelle de bouffons" ["War of the
Comic Actors"], see Cuillé, 2006.) For Mozart's librettist, words
(at least in opera) are all-important (Hilferty, 2006, p. 13). In the
nineteenth century, in his comments on music, E. T. A. Hoffmann
threw his authority, retrospectively, behind Gluck and the "Prima
la musica" ["music first"] school. (For an exhaustive study of the
relation between music and language in Hoffmann, see Hamilton,
2008.) Hoffmann (n.d., p. 41) favors purely instrumental, non-
programmatic music that takes us to a country "die nichts gemein
hat mit der äussern Sinnenwelt, die ihn [den Menschen] umgibt,
und in der er alle bestimmten Gefühle zurücklässt, um sich einer
unaussprechlichen Sehnsucht hinzugeben" ["which has nothing
in common with the external world of the senses that surrounds

us, a country where we leave behind all specifiable feelings, abandoning ourselves to an inexpressible longing"]. With respect to song, in which words that refer to specific affects are unavoidable, Hoffmann says that music confers a precious and noble quality on every emotional detail. He offers an extraordinarily precise formulation of a complex problem: through music, as, for instance, in opera, "selbst das im Leben Empfundene führt uns hinaus aus dem Leben in das Reich des Unendlichen" ["even life-experiences are made to lead us out of life, into the kingdom of the infinite"] (p. 42). I believe that this argument resembles the one that I offer on behalf of metaphor in a previous work (1991): metaphor is an imperfect image, drawn from life, seen through a perfect one (see also Chapter 5, Section 3a, below).

In Richard Strauss' *Capriccio* [*Caprice*], a twentieth-century treatment of the above-mentioned conflict over Gluck's operatic innovations, many of the topics that I discuss below are debated on stage. (Strauss' opera also draws on the still earlier "Querelle des bouffons," mentioned above, which is about similar issues.) In particular, some of my own arguments for the primacy of music, perhaps in human evolution, but certainly in the word-music combination, are voiced by Flamand, the musician. The opera tells us that the cry of pain (i.e., the precursor of musical expression) precedes language; the words to a song are unimportant; the words are usually worse than the music ("Fugue"); and, in any case, words are easily overwhelmed by music (scene 6). It is hard to know how to interpret Schoenberg's surprising remark that, for a truly musical listener, the text of *Pierrot lunaire* would be of no importance (Dunsby, 1992, p. 44).

5. IN THE WORD-MUSIC COMBINATION, MUSIC OVERRIDES WORDS

In a spirit consistent with Flamand's (and probably Strauss'), I would repeat the contention that, when we fit music to words, the music immediately overwhelms the inherent musicality of the language. I will speak later of the common situation in which good music redeems weak words. (The Canadian correspondent and musician, Steven Freygood, was quoting an unknown source when he remarked, "If it's too silly to be said, it can always be sung"; Mr. Christopher Chung has since kindly identified several possible sources for the quotation, among them Joseph Addison's *Spectator* [(1711) 1836, vol. 1, p. 42]: "That nothing is capable of being well set to music, that is not nonsense.") It is also possible for weak music to weaken a strong text. An example of a good poem with a mediocre musical setting is the familiar "Drink to me only with thine Eyes" by Ben Jonson, in which a lovely opening line is gradually turned by the laboring melody into something mawkish and insignificant. Similarly, in Goethe's lyric "Das Fischermädchen" ["The Fisher-Girl"], which Schubert set to music, a silly flourish in the tune at the end of each stanza destroys the sobering undertones in the playful text. Professor James Currie has reminded me that Saint Augustine (1909) was already worried by the ease with which the musical setting of the Psalms could make him forget about the words; Augustine (*Confessions*, Book 10, p. 193) cites Athanasius to the same effect. Storr (1992, p. 22) cites the same passage, and Professor Currie has told me that this continued to be an issue in medieval music. In the nineteenth century, the religious poet Gerard Manley Hopkins called only on "Elected Silence" to sing to him: "Shape nothing, lips; be lovely-dumb."

Although music and language may have arisen together, music undoubtedly takes precedence when they are combined, overriding language not only with respect to sound, but even with respect to meaning. (The composer Philip Glass [2006] has said the same about the relation of music to film.) Even more puzzling is the fact that one can have radically different musical settings for the same text or, for that matter, vice versa. For instance, "The Star-Spangled Banner" was, notoriously, grafted onto the tune of a British drinking song. Neither text is in itself of interest; whatever flavor either text may have is provided by the melody, which somehow contrives to flesh out and expand the meaning of the insignificant lyrics, lending some impetus to the different words in each case. ("Parody" is the technical term for the use of the same music with different texts.)

Some passages contain words that seem merely to be excuses for musical divagations, cadenzas, or melismas, in which meaning is not quite left behind, but pretty nearly: for whatever specifiable meaning "hallelujah" has as Handel puts it through its paces, the word might almost as well be a string of nonsense syllables. I am reminded of what has been said of the global aphasias: "there is no more than . . . a recurring stereotypic utterance," but "the recurring utterance may be repeated rapidly in a richly inflected manner that suggests fluent output if only it could be comprehended. . . . The mechanism of this richly inflected stereotype is unknown" (Alexander, 1997, p. 139). One might also think of the Lapp (Sami) songs, which consist of nonsense syllables that nevertheless "always carry personal meaning" for the singer and are identified with a specific topic (Gray et al., 2001, p. 53; on nonsense songs, and on the relation of music to language, see also Williams, 1980, pp. 44–46). Anne Fernald (1989) identifies four melodic profiles in "motherese," the semi-nonsensical form of communication that mothers use with their infants. These melodic lines are found worldwide, are independent of any particular language, and are therefore apparently innate. It is tempting to think that variations of this universally intelligible musical language are embedded in what we recognize more formally as music.

As I suggest above, a wide range of moods and attitudes can be attributed to the same words by music. (For a simple example, compare Schubert's "Ode to Joy" with Beethoven's, or all the various settings of the Mass.) It is as if music can run away with meaning whenever it chooses to do so, as if, in such cases, the words were no longer the carriers of the meaning, but only supported and expressed that of the music. By its mere physicality, music functions to narrow and specify meaning. As the Greeks were well aware, it moves language into the performative, or enactive, mode. When

Schubert's Erlkönig says, "Und bist du nicht willig, so brauch ich Gewalt!" ["And if you won't come, I'll take you by force!"], we have no choice but to follow. When Goethe's Erlkönig says the same thing, we can follow or stay where we are, take it or leave it. (For some interesting but enigmatic remarks on the relation of text to music in Schubert, see Schoenberg, 1975, pp. 144–145.)

There is also the fact (again, implied above) that music can use trite, utterly conventional words, transfigure them, and invest them with all sorts of meaning (as is true, again, in most of the Schubert songs), whereas good words cannot rescue bad music from triviality. Even in those (surprisingly rare) instances in which the verbal text is in itself valuable, the words have to renounce their own phonetic identity, allow themselves to be stripped of their acoustic DNA, so to speak, and let the music fill them with its own purpose. In the ballad form, in which the words are often of uneven quality, but in which the tune remains the same, a good melody lends beauty when the words are weak, sustains and exalts when the words are strong. Once in motion, music sweeps up the words and carries them along with it, in a state somewhere between sleeping and waking, a state where the normal declarative forms of language that are integral to waking consciousness are no longer operative. Paustovsky (1986, p. 169) states that "it's one thing to say something, but if you sing it that's quite a different matter." Perhaps, to the three familiar conditions of wakefulness, NREM sleep, and REM sleep, we need to add a fourth: the condition of music (see Sections 6 and 7, below).

The most imperious assertion that I know of music's power over words appears in Toni Morrison's *Beloved* (2004). In the novel, a group of women gathers outside Sethe's house to try to rid her of the incubus of Beloved, the baby whose throat she had cut in order to protect her from slave-hunters, and who has come back to haunt her. As the crowd chants, "the voices of women searched for the right combination, the key, the code, *the sound that broke the back of words*" (p. 308, my italics). The quieting of the ghost can only be achieved by the destruction of language, by getting back to a unity in which violence cannot happen. That unity had been fatally compromised by language. Language is named as the arch-enemy, the root of all evil. One is reminded of Walter Benjamin's ([1916] 1979, p. 327) longing for a prelinguistic, prelapsarian state: "the Fall marks the birth of the human word." Morrison's accusation, like the ghostly child, is pregnant with meaning. Only by singing can we get back to a time before words; moreover, only by singing, by finding the sound that breaks "the back of words," can we undo the damage that language has done to us. To be baptized once more, we have to be washed clean of language. By music.

In this context, it seems fair to revisit the question about the historical precedence of music over language. As Flamand, in Strauss' *Capriccio*, points out, some sort of cry, even if it were only the cry of pain, must have preceded language; whether that cry would ipso facto have qualified as music is another question. But to confront the "origin of language" question directly is for braver minds than mine. Nietzsche assumes, on philosophical grounds, the primacy of music (Kofman, 1971, pp. 783–788). Professor Jason Brown has reminded me that Wagner followed Darwin in assuming that music preceded language. I am interested in the issue of the primacy of language or music as it arises in sleep studies, which show that the two functions behave quite differently in dreams. It is obvious that music and language are interrelated, and I would suggest that music, which specifies mood and emotion more closely than language alone can do, may have been an aspect of early language. (It may be worth noting that the right hemisphere, which is more biased toward music than is the left hemisphere—though Levitin [2006, p. 9] disagrees—also houses certain verbal capacities that may represent the vestiges of a proto-language [Baynes and Gazzaniga, 2005]. The cry or exclamation, as a manifestation of emotion, is probably also primarily a right hemisphere function.) In sum, although I frequently waver on this issue, I am inclined to believe, like other proponents of the "joint precursor" hypothesis (S. Brown, 2000, and Molino, 2000), that music is coeval with language, though perhaps in some pre-rhythmic form it may have preceded language. (Jourdain [1997, pp. 306–307] argues that rhythm is not fundamental to music; see also Patel, 2008, pp. 97–98.) If music did precede language, it could not have been in the form of song as we now understand it, since for us song generally involves words.

6. MUSIC MAY WORK TO THE ADVANTAGE OF THE RIGHT HEMISPHERE

Another important issue in the characterization of music is music's apparent affinity with the right hemisphere, which is sometimes said to become more active during sleep (Moscovitch, 1973). Now, I think it is fair to say that music is not just something produced deliberately; like other creative experiences, it is something that lies in wait for one rather than something one pursues. The music that really counts is the music of which one can say that one literally does not know where it comes from. If we understand music to be an affirmation of the wholeness of our being rather than some minor activity, it would be demeaning to conceive of it as mere therapy. (Still, Levitin [2006, p. 187] points to the fact that music raises dopamine levels in the nucleus accumbens, thereby improving our mood.) Nevertheless, to embark on a brief, purely speculative detour, and allowing for all the complexities and uncertainties that surround the issue of lateralization (Harrington, 1987; Efron, 1990; Uttal, 2001; and Edelman, 2004), especially with respect to music (Tramo, 2001, and Warren, 2004), is it possible that the right hemisphere *needs* music, rather than that the right hemisphere is merely better at processing music? It is often said that the right hemisphere has depressive tendencies (Davidson, 1995; McNamara et al., 2002, pp. 131–132; Harmon-Jones, 2003; and Rotenberg, 2004). Could music be a tonic, or a stabilizing force, for the right hemisphere—even in sleep? If so, it would be easier to understand why "Our sweetest songs are those that tell of saddest thought" (Shelley [1820] 1956, "To a Skylark," line 90)—songs in which the right hemisphere, conduit for sorrow, has been rescued from a deeper discouragement by the musical tone. Music would, then, be a "treatment" for the right hemisphere, reviving and enlisting its emotional, metaphorical, and cognitive powers on behalf of the organism, and restoring its functional unity. As homeostatic systems that are miraculously but temporarily balanced and constantly threatened by disequilibrium, we need all the help we can get in stabilizing our permanently precarious position. (For a history of ideas about the healing power of music, see G. S. Rousseau, 2004, pp. 46–54. On the neurochemistry of our response to sad music, see Levitin, 2008, p. 133.)

7. THE COMBINATION OF WORDS AND MUSIC MAY CONSTITUTE A SEPARATE MODULE

The Brain Music Therapy program at Toronto Western Hospital uses music adapted to an individual's EEG (Kayumov et al., 2002). At this point, some mention of Melodic Intonation Therapy (MIT) (brought to my attention by Dr. Alan Lockwood of The State University of New York) seems unavoidable, especially since it involves questions of brain laterality. MIT was developed as a result of the observation (mentioned above) that even some global aphasics can sing words that they cannot speak (Geschwind, 1965, p. 633, and Klawans, 2000, pp. 132–140). A treatment for global aphasia was consequently developed, in which words were sung (rather than spoken) with drawn-out syllables and a contrastive pitch. It was thought that this procedure helped to bring language functions, which had been abnormally relocated to the right hemisphere, back to the left hemisphere, thereby assisting in the process of recovery from aphasia (Belin et al., 1996, p. 1509). This theory has been disputed (Hébert et al., 2003), and the recommended procedures have been modified (Boucher et al., 2001), but it was developed because the combination of language and music seems to constitute a distinct function, possibly served by its own neural array, with a nature and an importance of its own. (Katie Overy of Edinburgh University has told me that both the theory and the technique of MIT are still very much in flux.)

126   Since other apparently unrelated modalities, from imagery and abstraction to sound and color, can become linked in either transient or fixed relationships (Hunt, 1995, pp. 154–160 and *passim* on cross-modal "translation"; also Stein, Wallace, and Stanford, 2000, on "Merging Sensory Signals"), why not music and language? As Ilona Roth says, "It is possible that the complementary mechanisms for music in the absence of language, or vice versa, are not the same as those which exist when both language and music are present and function normally" (quoted in Mithen, 2005, p. 297, fn. 1). I have suggested that music, by itself, creates a distinct state of consciousness; music with words, in turn, is not merely an enhancement of language, it is a replacement of language by another "module" of our communicative system.

An article by Sylvie Hébert and Isabelle Peretz (2001) appears to cast doubt on this conclusion. These authors maintain that the relation of words to music in a song is one of association, not of integration: if text and music were actually integrated, "the mental code for a song would not retain separate verbal and musical codes, but rather some unique, mixed code" (p. 170). Hébert and Peretz observe that a brain-damaged patient who had difficulty in processing music had no difficulty in recognizing the text of a song, even when presented together with an irrelevant tune; from this and other ingeniously elicited responses, they conclude that text and tune draw on "independent processing resources" (p. 174). In a more recent article, "Singing in the Brain: Insights from Cognitive Neuropsychology," Isabelle Peretz et al. (2004, p. 386) dismiss previous studies as inadequate and argue once more, on the basis of a new case, "for the complete separation of music and language processors." Similarly, Robert Zatorre has pointed out to me in conversation that we can still recognize the words of a song when we hear them without the music. In this instance, my reply would be that we can also recognize a tree by its mere outline on paper, but that we respond quite differently to that shape as part of the living tree. (A related example would be our response to rhyme, which, when understood in the context of a poem, is no longer just the matching or repetition of similar sounds.)

8. THE FACT THAT THE RIGHT HEMISPHERE IS NOT COMMITTED TO DISCURSIVE LANGUAGE MAY HAVE SOMETHING TO DO WITH MUSIC'S DOMINANCE OVER WORDS

In contrast to Peretz et al., Jeffries, Fritz, and Braun (2003), in an article cited by Peretz et al. themselves (2004), state that "functionally distinct networks within the right hemisphere may underlie production of words in melody"(p. 754), and that this fact may explain the very facilitation of speech by singing that Peretz et al. deny. Schön, Gordon, and Besson (2005, p. 75), though pointing out some limitations of the Jeffries et al. (2003) study, seem to come to much the same conclusion. Steinke et al. (2001, p. 428) suggest that "songs (with lyrics) and instrumental melodies are processed and stored in different ways." In an earlier study, "Singing with and without Words: Hemispheric Asymmetries in Motor Control," Andrea Cadalbert et al. (1994) defend quite emphatically the special role of the right hemisphere in singing with words. I believe that the findings of Lauren Stewart et al. (2001) also provide more support for my position than they do for that of Peretz et al. (2004). The title of their article speaks for itself: "Transcranial Magnetic Stimulation Produces Speech Arrest but not Song Arrest." Since this is the case, we are again obliged to consider seriously the likelihood that "there may be something special about word production while singing" (Stewart et al., 2001, p. 434). As I, too, have been saying about the primacy of music in word-music combinations, "melodic production may just be more robust than speech production" (Stewart et al., 2001, p. 434; unlike that of dream speech, the tendency of dream music to persist after one wakes also becomes easier to understand).

Brian Kolb and Ian Q. Whishaw (2001, p. 341) provide further strong evidence of a right hemisphere bias for song: they report the case of a patient whose left hemisphere (including Broca's and Wernicke's areas) died, but who was still able to sing a large number of songs that she had learned before the onset of her illness. In fact, reports of a continued ability to sing with words in cases of global aphasia go back as far as 1745 (Shlain, 1991, pp. 396–397, and Benton and Joynt, 1960, pp. 211–212). It has also been shown that new songs with words can be taught to a global aphasic (Klawans, 2000, pp. 137–138). (Patel [2003] has begun to address the complex problem of the partial overlap between language and music.)

It is probably futile to seek a physiological explanation for a problem in aesthetics, but one might at least speculate that, if the word-music combination operates under the aegis of the right hemisphere, the dominance of music over language might be explained simply by the characteristics of the right hemisphere, particularly by the fact that the right hemisphere is usually dominant for music, but is limited in language. (See Stephen La Berge [1992, pp. 300–301] on right hemisphere dominance in singing even during lucid dreaming.) The subordination of language in the right hemisphere would then be reflected in the subordination of language in the word-music combination. At times, music even seems to be pleading for an expression that is denied it; at other times it falls back upon gibberish, or upon a forgotten or an incomprehensible language, in its futile attempt at fulfillment (Wordsworth ["The Solitary Reaper," (1808) 1965, line 17]: "Will no one tell me what she sings?" Coleridge ["Kubla Khan," (1816) 1957, lines 42–43]: "Could I revive within me / Her symphony and song"). If, even in poetry, language is "forced into an interaction with" the right hemisphere (Turner and Pöppel, 1988, p. 83), in song it must surrender its autonomy almost entirely. In a song, the right hemisphere finds itself with words on its hands that it cannot have generated itself, and it is called upon to do something with those (borrowed) words. What it will do, naturally, is impose its coloring on the text through volume, stress, pitch, and whatever other pigments on the musical palette are available to it. What we get in the end, when we set words to music, is something like language, but it must be a different kind of language. When we hear words combined with music, we do not assimilate them as mere propositional language or even as normal poetry. For better or for worse, music invests them with a new quality. They assume the particular meaning conferred on them by the music. In

the famous lament of Dido in Henry Purcell's *Dido and Aeneas*,　
the commonplace words, "Remember me, but ah! forget my fate,"
become unforgettable only because of the music.

Dr. Anne Blood of the Harvard Medical School has suggested
to me that music dominates words simply because music is more
emotional than words. Of course, it is very difficult to specify what
extramusical emotions, if any, music embodies (cf. Wallin, 1991,
pp. 485–491). Theories of emotion still seem to be rudimentary: the
number of emotions, their interrelations, and their gradations is infinite, and the fact that we do not have a name for a certain kind of
emotion does not mean that it does not exist. Though music clearly
produces an emotional effect, it would be a challenge to name the
emotions "expressed" by a Beethoven quartet. (David Huron [2006],
to whose work I was referred by Jason Warren, makes an effort to
deal with this difficult subject.) Given the present incomplete state
of the theory of emotion, Dr. Blood's argument that music is more
powerful than words because it is entirely physical, whereas language always contains an element of abstraction, is more helpful,
though that fact in itself should not necessarily lead to the subordination of meaning. (Illustrating a text with "physical" pictures,
which is a process that bears some resemblance to illustrating a text
with music, seems to replace the text by total substitution rather
than simply absorb it, as music does.) I venture, once more, the
conjecture that right hemisphere dominance, perhaps in conjunc-
tion with the greater physicality of music, accounts for the ease
with which music can manipulate words.

It is possible to set this issue in a wider context: our brain may
not be able to tell us what we can think, but it may be able to tell us
what we cannot think. Some familiar cases include the difficulty
of parsing "onion" sentences (Pinker, 1995, pp. 204–207) and the
difficulty of remembering more than seven (or, possibly, nine; see
Levitin, 2006, p. 214) items at once (G. A. Miller, 1962, pp. 47–48).
Perhaps the brain has also set limits on our ability to maintain the
semantic independence and salience of a verbal text in a musical
setting. Language pays a heavy price for admission to music, though

it can also become, by default, a vicarious voice for the language-poor right hemisphere.

In one sense, text and music are obviously separable: as I point out above, one can have different melodies for the same words, and vice versa. One may also forget the words to a song. The question is, rather, what is happening while words and music are together? The fact that a transformation has occurred becomes startlingly apparent when we try to separate the two after they have been conjoined: the music remains intact, but the text reverts to something completely different. (We do not often hear people walking around muttering opera libretti to themselves.) The combination of text and music results in a lopsided relationship. Who needs the words to Beethoven's "Ode to Joy," anyway? Besides, even if the words were very good on their own, no matter how moving, they would still have to be good in a manner completely different from the way in which they are good while they are acting at the behest of the music. To see how radical a change language undergoes when it enters the arena of music, one has only to conduct the simple experiment of composing a melody and then singing some favorite lines of poetry to it. One is likely to be amazed at the metamorphosis that the words undergo: they become virtually unrecognizable. What springs to mind for me is a passage from Paul Valéry's "Le Cimetière marin" ["The Graveyard by the Sea"] ([1920] 1957, lines 123–125):

> Zénon! Cruel Zénon! Zénon d'Élée!
> M'as-tu percé de cette flèche ailée
> Qui vibre, vole, et qui ne vole pas!
>
> [Zeno! Cruel Zeno! Zeno of Elea!
> Have you pierced me with that winged arrow
> That flutters, flies, but does not move at all!]

Not only does a ready-made melody destroy or, perhaps, replace the lines: one made to order is also likely to alter them irremediably. To reverse the experiment, remove the music from the following

lines of the "Marseillaise"—"Qu'un sang impur / Abreuve nos sil-
lons" ["That a foul blood / Slake our furrows"]—and recoil at the
horrifying thought that the gay little tune had concealed.

~~~~~

A postscript: Lying in bed, half asleep, waiting to get up, I felt different moods, groups of thoughts, shifting through my mind—perhaps every two or three minutes, though I do not really know—like
packets, discrete quanta of mood-thoughts. The defining feature
of these thoughts, what marked their boundaries decisively, was
that they were accompanied by different musics in my mind: the
musical accompaniment changed with the change of thought. As I
always have music running through my mind, the musical changes
produced clear markers.

What drives these changes? The eighteenth century thought
about this a lot—the idea of the mind as an Aeolian harp. What is
the wind? Chemical flows through the body, or the switching of
hormonal systems? What makes them change?

9. CONCLUSION

My purpose in exploring, at considerable length, the balance between music and language has been to account for the difference
in the behavior of the two functions during dream. Music in some
form appears to be a fundamental function, more deeply rooted
in our affective life and possibly more ancient than language. It is
also an art form, and as such it has an inner coherence that requires
it to maintain an integrated contour. It may be for these reasons
that music is more likely than our other mental functions to occur
unaltered during sleep, and to be recalled more consistently than
language after we awake.

The musical dream, then, affords us the possibility of engaging
major theoretical issues that would remain below the horizon of

132   investigation if we were to overlook it. The historical review of the literature with which I began this chapter reveals a steady curiosity and interest in the phenomenon, but it also reveals the fact that so far there has been no major effort to inquire into its significance or to take advantage of the opening it provides for insights into the music-language relation. It has been my hope to show that this topic remains fruitful for further consideration.

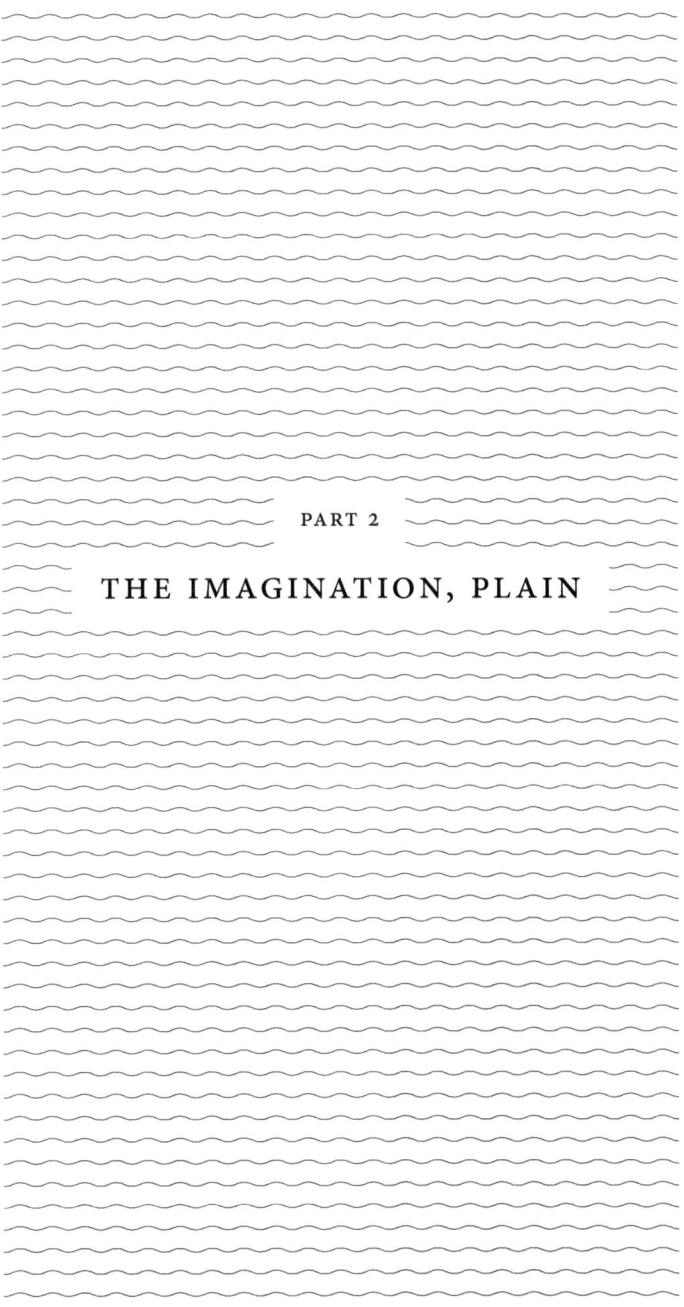

PART 2

THE IMAGINATION, PLAIN

5

The Three Fields

1. INTRODUCTION

The second part of this book offers a summary review of the differences between standard aesthetic and neuroscientific approaches to the arts. I present examples of "the imagination, plain," that is, essays written without recourse to neuroscience, from each of the three major fields of artistic expression: the visual, the verbal, and the musical. In each case, I emphasize the contrast with the neuroscientific approach. As Barbara Herrnstein Smith (2005, pp. 110–112) shows, though, any attempt to organize the comparison of these two approaches in a strictly systematic manner is fraught with difficulties. A discerning reader may also find some structural irregularities in the remainder of this book.

The fact is that four chapters in the latter part of the book were omitted, with a view to making the work more coherent. In the process, certain links were elided. For instance, the material on tragedy (Chapter 6, Section 2, below), which may now appear somewhat isolated, was originally written as a postscript for a chapter on the topic of illness in Yiddish poetry. (This chapter was derived from one of my previous studies [2006b].) Likewise, three chapters on the imagination served as the basis for the theoretical commentary in Parts 2 and 3. (These can be found in Massey, 1977, 1991, and 1999.) Here again, for example, my remarks on metaphor in Section 3a, below, can be more easily understood against the background of Massey 1991 and 1999. On the whole, the material removed in the abridgement was originally intended to emphasize the differences between the humanistic and the neuroscientific approaches, and to define the areas in which the scientific method is least useful. It is hoped that enough of the original structure remains so that the reader will be able to follow this chapter's argument without difficulty.

2. THE VISUAL ARTS

It is worth recalling that some of the relevant comparisons between the aesthetic and neuroscientific approaches arise quite naturally with respect to cubism (see Chapter 2, Section 1, above). The disorientation and reorientation that cubism requires has a good deal to do with the effects that Margaret Livingstone (2002) analyzes in her work on color-luminance imbalance and pointillist illusions of motion in motionless paintings. In each case, the tired habits of the eye, which ceases to see anything if it sees the same thing long enough, must be overcome; something is made strange so that, when we recognize it again, we are compelled to see it as new (pp. 150–153). This technique is the familiar "making strange," the Russian "os-tranienie" or the German "Verfremdungseffekt." A simple version of it can be found in Matisse's "Femme au chapeau" ["Woman with a Hat"]—which is composed of patches of color that have no resemblance to the colors of the subject (p. 134)—or in Derain's portrait of Matisse (p. 136). Of course, the decomposition-recomposition strategy is taken to an extreme in Chuck Close's self-portraits.

Livingstone grounds all her analyses of painterly effects in detailed neuroanatomical information. For instance, she suggests that cubism may work because of "neurons that will respond exclusively to a particular object, at various viewing angles. This means that some memory templates in our brains are view invariant; that is, that you can recognize an object or person seen from any angle" (p. 77). The ensuing idea "that cubism is pleasing because it resonates with a view-invariant part of our memory system" (p. 77) is intriguing, but it is hard to tell how far it actually takes us. Why the selective activation of a group of space-recognition neurons should be pleasing is not clear; perhaps the pleasure lies, rather, in our having solved a visual puzzle (cf. Ramachandran and Hirstein, 1999, pp. 20–22). Ramachandran and Hirstein likewise speak of the possibility that "some types of art such as cubism are activating brain mechanisms in such a way as to tap into or even caricature certain innate form primitives [cf. Lipps, 1891] which we do not yet fully understand. At present we have no idea what the 'form primitives' used by the human visual pathways are" (p. 20). Of course, none of these suggestions brings us closer to understanding why we prefer some cubist works to others, why those preferences may change over time, or why some viewers do not take any particular pleasure in cubist techniques to begin with.

The difficulty with reducing our responses to art to neurological patterns becomes apparent in our reaction to individual works, as when Livingstone resorts to the concept of equiluminance in order to account for her experience of jitteriness when viewing Mondrian's "Broadway Boogie Woogie"; for me, however, the experience is caused by a sense of structural imbalance (see Chapter 2, Section 2, above). My own response to the painting is entirely intuitive and impressionistic; but, even if it were translated into neurological terms (which would not be difficult to do), it would not alter the fact that some subjective criterion is the final arbiter in our decisions. Two viewers of the same picture may notice and emphasize different things about it, whatever their interpretive platforms.

The technical discussions of Monet in Livingstone's book are also connected to the above-mentioned "making strange" paradigm that was influential throughout twentieth-century aesthetics, though less so now that the "strange" has become the commonplace. Monet criticism has been said to fall into three camps: the ahistorical or formalist variety, the historical, and the psychological-biographical. Of these, the first two are not readily accessible to neuroscientific discussion. It would be hard to see, for instance, how Octave Mirbeau's oft-quoted comment on Monet's haystacks—that they constitute "the revelation of the states of consciousness of the planet" (quoted in Stuckey, 1993, p. 4)—could be translated into neurological terms. The biographical question regarding Monet's cataracts, and the way in which physical problems (cf., again, Chuck Close) relate to the aesthetic, is again a subject for a separate kind of inquiry. There is, however, one phenomenon that I discuss in Chapter 2, Section 2, above, that lends itself readily to an illustration of the contrast between aesthetics and neuroscience. This phenomenon is the preference for "parentheses" over "hourglass" configurations noted by Ramachandran and Hirstein: the preference for structures that offer closure over those that open on an undefined space.

I have run across a nineteenth-century article on a closely related subject: the preference for concave over convex forms (i.e., for an enclosing shape over its opposite), couched in a traditional vocabulary that contrasts sharply with Ramachandran and Hirstein's neurological analysis of the phenomenon. (I could have used Elaine Scarry's [2004, pp. 251–254] rhapsody on concavity, but it is more complex and more literary, and therefore less manageable for comparison than is the simpler nineteenth-century example.) I believe that setting the two explanations side by side gives one a good sense of their differences as they approach a related problem in visual aesthetics. Full quotations of relevant passages will make the contrasts clear. Although it is more recent than either of the articles I just mentioned, the article by Bar and Neta (2007) on the human preference for curved lines or shapes seems to fall somewhere between these two approaches.

The nineteenth-century article in question is an 1878 review of Gustav Theodore Fechner's 1876 work titled *Vorschule der Aesthetik* [*Introduction to Aesthetics*]. The review was written by Calame, and it appeared in the *Revue philosophique*, edited by Ribot. Calame begins by declaring, "It seems that our sight is repelled by convexity and attracted by concavity," and that "we are repelled by a convex surface; a concave surface attracts us and appears ready to take us in" (p. 181). He goes on to state that

> the back is arched in order to deflect a blow. The chest of a
> strong man is thrust forward to meet that which it seeks to

repel. The fist closes against the enemy to frighten and to
strike him. Horses form a circle to defend themselves against
wolves. The bridge arches over a torrent to provide safety
for the traveler. A bomb bounces off the cupola of a dome.
Rain runs off an open umbrella. . . . An open door that leads
into a house: doesn't it invite one to enter? Similar experiences, repeated daily, end up leaving impressions that are
associated with convexity and concavity, and, according to
the circumstances, they take on an agreeable or a disagreeable character. One takes pleasure in looking up at the sky
or into the vault of a cathedral; the soul is, so to speak, lifted
with the gaze. Try to picture the unimaginable: convex skies
or ceilings; the impression would be painful, crushing, as
if they sought to bury us in the earth. That is why garlands
suspended from one house to another, over streets, during
holidays, look bad. (p. 181)

Calame goes on to discuss the shape of sofas: the upholstery, though
convex, is inviting, because we are aware of its elasticity and of the
opportunity it provides us of creating a concavity in it (p. 181). (It
is amusing to find Bar and Neta [2007] also using a sofa as their
example: "the amygdala . . . is significantly more active for every-
day sharp objects [e.g., a sofa with sharp corners] compared with
their curved contour counterparts" [p. 2191].)

The passage from Ramachandran and Hirstein is much more
difficult to excerpt, since the explanation it offers for our preference for closed forms is embedded in, and dependent on, a larger
explanatory scheme. The brain needs to exercise "binding" in order
to enable us to recognize objects in a "noisy" environment: to keep
ourselves safe, or to get what we want, we have to identify what may
be hiding out there; we have to put clues together.

[T]here may be direct links in the brain between the processes
that discover such correlations and the limbic areas which
give rise to the pleasurable "rewarding" sensations associated
with "feature binding." . . . [P]roximity along different feature

dimensions may be useful for perceptual grouping and "binding" of features that are similar within that dimension. . . .

The visual system is often called upon . . . to defeat camouflage—and this might be easier to accomplish if a limbic "reinforcement" signal is not only fed back to early vision once an object has been completely identified, but is evoked at each and every stage in processing as soon as a partial "consistency" and binding is achieved.

. . . [O]ut of a temporary binding a signal [is] sent to the limbic system to reinforce the binding, and this is one source of the aesthetic experience. For example, in . . . [the parentheses vs. hourglass example], there are two possible stable organizations, one with hourglasses, and one with closure and most people find the latter organization more pleasing than the former because the limbic activation is stronger with this closure-based object-like perception. (pp. 22–24)

The first thing that one notices about the two accounts is that they both explain the advantage of the enclosed form as a function of safety and of the pleasure of feeling secure. Both also reason backward rather than (predictively) forward: only if one has first observed the preference for enclosed forms can one search for an explanation for that preference. The main difference between the authors lies in the explanatory material on which they draw. Calame relies on an associative principle whereby we identify concave forms from our ordinary experience with more satisfactory situations than convex forms (breasts are noticeably lacking!); Ramachandran and Hirstein rely on unconscious processes, an inner calculus of advantage that the brain performs on our behalf. Their argument has larger implications than does Calame's, connecting aesthetic "closure" with many other versions of that general psychological function; however, Calame did not set out to propose a broader thesis.

As far as the usefulness of either argument for aesthetic judgment or generalization is concerned, the same objection may be raised.

As I point out above, closed or semi-enclosed forms are salient and ubiquitous in the visual arts, but we can never decide on the value of a particular work on the basis of its having used, or failed to use, closed forms as a major feature of its visual narrative. Then again, in some cases, closed forms may be disagreeable, even hideous, as in the bursting bodies of the German expressionist caricaturists George Grosz or Max Beckmann. Finally, the problem with the neurological explanation of this feature of our perceptive response is the same problem that dogs almost all such explanations: it really has nothing to do with aesthetics as such.

Ramachandran and Hirstein begin their exposition with a visual puzzle: a Dalmatian dog has to be picked out from a bewildering flurry of dots. Aha! We have discovered the dog, and we are happy. That happiness is somehow similar to the happiness we experience when we find parentheses among the bewildering hourglasses. That happiness is, in turn, somehow similar to the happiness we experience when we view the (scarcely camouflaged!) responsive curves in the bodies of the sculptured Indian goddesses that figure largely among Ramachandran and Hirstein's illustrations. But in proposing that last step, the authors confuse categories: a psychological category is not ipso facto an aesthetic category, as even Ramachandran and Hirstein begin to acknowledge in their subsequent elaboration of their thesis. The "pleasure" in viewing a work of art is not the same as the pleasure that one has in an ordinary experience. In fact, the encounter with a work of art may not have anything to do with "pleasure" in the first place. Does one regard the Kenwood Rembrandt self-portrait (of which more below) with pleasure?

At least Calame has the merit of drawing his explanatory materials from the same drawer, so to speak, from which artists take their materials: from life itself. The moment one invokes the vocabulary of the brain, one creates a gulf that may be impossible to cross. I quote Taine (1877, p. 9): "When thought takes on the appearance of physiology, it does so because it has been translated into a language that is foreign to it, a language in which it necessarily takes on a character not its own." ["Quand elle [la pensée] revêt l'apparence physiologique, c'est qu'on la traduit dans une langue étrangère, où, forcément, elle revêt un caractère qui n'est pas le sien."] On the one hand, when one speaks of unconscious neural processes as though they translated spontaneously into a vocabulary continuous with our own, one is assuming a great deal. On the other hand, one may argue that the attraction of closure is more of a reflex than it is a thought, and so is fair game for cognitive neuroscientists. Still, a universal and unconscious neural process, such as the overriding appeal of closure in art, entails an automaticity that has hitherto been regarded as incompatible with the aesthetic. But that may be the very presumption in question, or at stake, in neuroaesthetics, and one could be accused of engaging in a circular argument if one were to reject it out of hand.

To set this question aside for the present, I would like to refer briefly to the recent article on contours that I mention above (Bar and Neta, 2007). This article contrasts responses not to concave and convex surfaces, but to curved and angular contours, and it does not distinguish between convex and concave curves. Nevertheless, the two topics are related, and it is interesting to compare the approaches of the various commentators on these issues. Calame's approach is entirely phenomenological; Ramachandran and Hirstein's is strictly neurological; and Bar and Neta's seems to fall somewhere in between, since, though they work with neurological data, they offer the hypothesis that sharp contours are intrinsically threatening. All three, though, seem to operate within the polarity of danger and security.

It is rather surprising that none of these commentators mentions the extensive attention paid in the eighteenth century to their subjects—for instance, in William Hogarth's work on the line of beauty. It would also have been a good idea for them to recall Erasmus Darwin's opinion that we enjoy curves simply because they remind us of the maternal breast. Not only does this explanation have the benefit of common sense, but it also serves as a bridge to the aesthetic, since Darwin proposed that it explains the origin of the idea of beauty itself. From the point of view of aesthetics, though, the following problem still arises: no matter how many neurobio- logical universals we may identify as contributing to our general response to art, they never determine our particular response to any particular work.

3. LITERATURE AND METAPHOR

In my comparison of the aesthetic with the neuroscientific approaches to the visual arts, I used, as my example, not two different treatments of a single artistic work, but two treatments of one problem in visual aesthetics. In dealing with literature, though, I will use the style of a single author as my example and attempt to present two perspectives on his work. Before proceeding, though, I wish to return to the subject of metaphor, since it is a general theme that underlies all discussions of literature.

a. Metaphor

In Chapter 3, Section 3, I pointed out the resemblance of the metaphoric process to the well-known psychological situation in which "the appeal of the rare" comes into play. I also suggested that oxytocin might have some role in the pleasure associated with the experience of metaphor. My purpose was to find some physical process or circumstance that could be correlated, as closely as possible, with the event called metaphor.

More specifically, I would like to point out the fact that most metaphors involve visual elements. This fact may facilitate the transition from Section 2 ("The Visual Arts"), above, to the subject of literature. Thus, for instance, I see and name something that is (say, a swinging frond of willow) through something that is not (say, the slow swinging of a dancing elephant's feet), and a possible *literary* metaphor emerges. To proceed to a thesis or generalization concerning metaphor, in a previous essay (1991, p. 29), I suggest that a metaphor is "an imperfect image seen through a perfect one." To

144 illustrate this concept, I use the example of a flight of gulls cross-
ing, at a distance and very slowly, a broad, shallow subarctic valley
on the island of Miquelon, and settling, one above another, on the
opposite green slope. This is an "imperfect image," which I saw
through the "perfect image" of a series of marble steps being built
by invisible hands across the valley—an image drawn from the
storehouse of perfect images that each of us harbors.

The thesis embodied in this example is obviously bold, if not
far-fetched, and it requires an elaborate, though perhaps ultimately
futile, defense. It is also very difficult to envisage a neurological version of the same scenario, though one might attempt to pursue the
possibility. Perhaps one could find the lair in the brain to which the
"perfect images," once formed, retreat; after all, they have a certain
affinity with dream images, the formation of which is considered a
legitimate field of scientific investigation. One might even want to
think about the neural representation of those metaphors sometimes called "concrete universals": is Achilles as lion a general, distributed representation that is somehow united, with peculiar force,
under the influence of emotions generated in the amygdala, until it
settles into a cliché, as one of the forms of memory associated with
the hippocampus? At the very least, metaphor is clearly connected
to a form of memory that might be explored neurologically. (For an
attempt to produce a model of metaphor in a connectionist system,
see Martindale, 2007, p. 188.)

Whether or not one could relate this thesis on metaphor to neurology, there are immense differences between the way I arrived at
it and the way a scientist would analyze it. Both procedures must be
scrupulous, though, each in its own way. The scruple in the "humanist" account lies in surrendering to a subjective experience without
withholding anything: being absolutely faithful to the phenomenology of the experience. On the other hand, my hypothesis, or,
perhaps, merely ignorant assumption, about the scientist's scruple
is that it consists in a ceaseless wily watchfulness for unforeseen
alignments in one's objective materials, always remaining available
to the unexpected. Perhaps, another kind of metaphoric experience,
also matching the expected and the unexpected.

Another feature of the humanist hypothesis distinguishes it from the scientific one. If the generalizations distilled from my experience on Miquelon have any worth, it is because something of that landscape still clings to them: what comes from those experiences is not a naked abstraction. (Cf. Badiou [2003] on identity formation as a commitment and fidelity to a specific event.) In a sense, these generalizations belong only to that situation—to the situation in which they were born—and they carry its atmosphere with them forever. In other words, as generalizations, they are still metaphors. For this reason, when one tries to apply humanistic generalizations retrospectively, to previous periods or movements, they often appear unnatural or inappropriate. They are basically historical, that is, outgrowths of their own time and place, solutions to the problems of a specific and, later, a personal or historical moment. So it is not hard to understand our uneasiness when we hear about Laurence Sterne's "postmodern" style or about the Derridean "différance" in Wordsworth's endless postponements.

The scientific generalization also carries its history and the context of its origins with it, but, once launched, it has greater independence. Its value is no longer measured by the faithfulness with which it reflects the circumstances of its origin, and it should, I understand, apply equally to problems of the past and of the present.

Having disburdened myself of these thoughts on the subject of metaphor, which has obsessed philosophers of both art and science at least since Aristotle, I will go on to discuss a favorite author of mine from two contrasting, or, perhaps (and one must allow for the possibility) entirely incompatible points of view. It will soon be apparent that the topic of metaphor cannot easily be left behind.

b. Patrick Fermor

Recently I read a book that I told my friends, only half-facetiously, was the best book I had ever read. It was one of Patrick Fermor's travelogues (actually, two books): *A Time of Gifts* (1977) and *Between the Woods and the Water* (1987).

On hearing that I taught literature, a passing acquaintance once asked me what the book was that had given me the most pleasure. I bridled at the question. Reading isn't for "pleasure." The best books I could think of did not simply give me pleasure. In fact, many of them were painful to read, and even the pleasant ones did things that mattered to me a great deal more than did pleasure. (Of the latter group, one of my favorites is John Galt's *Annals of the Parish*.) Pleasure was inessential—at most, incidental.

"No," said my acquaintance, "I want to know what book gave you the most *pleasure*." I stumbled, stuttering, not knowing what to say. I would have been ashamed to say that I valued a book simply because it gave me pleasure. That is not what literature is for. The conversation trailed off unsatisfactorily.

Well, here was a book that gave me pleasure, and that I valued because it gave me pleasure. I positively wallowed in it, like the proverbial pig in a trough. I loved the basins full of exotic place names, the bewildering nomenclature of heraldry and architectural detail, the cascades of historical information. Twenty pages just about the seacoast of Bohemia! And the travelogue itself, ascending from the hospitable entertainment of the lowly traveler by mere counts to dukes to princes in a Des Esseintes-like narcissist progression (Huysmans, *À Rebours*) of increasing sophistication and opulence; the whole accompanied by landscape descriptions of unfailing novelty. This was Euphues redivivus, but the readers of Euphues never had it so good.

What was best of all about the book, though, even better than the wonderful metaphors that stud its pages, arriving with unforeseeable irregularity every couple of pages to buoy the narrative, was the extravagance of the language. With the word "extravagance," I do not refer to the narrative's lushness or fullness, to its astonishingly copious vocabulary, or to its deluge of odd facts overwhelming the reader. The extravagance lies in Fermor's attitude toward language as such. He treats the usual limits of language not as defining rules, but as suggestions. Language, as we usually think of it, consists of words or phrases that fit together in pre-established ways. But, when reading Fermor, we gradually realize that he is

driving an off-road vehicle that keeps to the linguistic highway only
so long as it happens to coincide with his purpose, but that takes
off across country whenever the highway is not the direct route to
his objective. (Shakespeare often does something similar, but so
casually that one does not notice how widely he expands language
[see, for example, *Troilus and Cressida*, 3.3.473; *Othello*, 4.1.39–40;
Antony and Cleopatra, 1.3.90; and *King Lear*, 4.6.102].) These divagations are not obtrusive or shocking, because they seem so natural: as I have said, they make one realize that language, properly
understood, is a handbook of suggestions rather than a road map
or an apparatus for getting one to a fixed destination, since there
is no way of knowing that destination in advance. The destination
is produced by the sentence as it moves onward, or, to paraphrase
Theodore Roethke ([1951] 1996, line 3 repeated), it learns by going
where it has to go.

I draw most of my examples from *Between the Woods and the
Water* rather than from *A Time of Gifts* because I happen to have
a copy of the former at hand, although the latter has more of the
kinds of passages and phrases I am thinking of: for instance (from
my notes), "The Nazi salute flickered about the pavement like a *tic
douloureux*" (p. 103); "The grey discs and pods of ice" (p. 115); at sun-
down "life ebbed with a shudder" (p. 118); "undrawn cover" (p. 199).
From *Between the Woods and the Water*: "The late-rising moon had
extinguished many of the stars. . . . The last quarter scattered the
woods with . . . the gleams of lit rock" (p. 193); the Czerna (black
river) "askim with wagtails . . . rushed on chopping and changing
down the gorge" (p. 208); "and in a little while a green-leaf song was
mounting the hillside and then slowly fading" (p. 209); "the other
shore would swing away into the distance with the stars spreading
like a momentary chart of the heavens" (p. 221). A clear example
of what I mean by Fermor's linguistic divagations: "Several of my
neighbours fingered strings of beads, but not in prayer; they spilt
them between their fingers at random intervals, as though to scan
their boundless leisure" (p. 229). There is no way to make that last
clause "scan" semantically, even when one realizes that Fermor is

using the word "scan" in its prosodic sense; scansion requires some regularity, whereas these beads are "scanned" at random intervals. A more casual sample: "the ship . . . expanding fast and enclosed in the confetti of gulls" (p. 237). These are almost, but not quite, standard metaphors; they emphasize that part of language that transcends mere comparison.

It is this realization that Fermor has taught me a crucial truth about the nature of language that I most value. I believe that this piece of instruction, the conferral of this major piece of knowledge—that meaning is created by the way words make contact with each other outside and beneath syntax (or, as Rachel Blau Du Plessis [2008, line 3] puts it, by the "noises darkening below syntax")—is a deeply moral act, despite the narcotic effect of Fermor's style. Apart from that, the sheer sensual pleasure of reading Fermor robs me of even the sense of guilt at my self-abandonment, at the abdication of personal responsibility and even personal identity that it creates. Fermor's style is a real siren call, a lotus-eater experience. I have always thought of reading as essentially immoral (see my comments on Chekhov's "Ward 6" in 1987a, pp. 70–78), but not immoral in this blatant and obvious way; there is almost always a fig leaf of morality that attempts to disguise its self-indulgence and narcissism. But carried to this extreme, and moving language into a mode that reveals the puerility of the restrictions usually imposed on it, Fermor's writing begins to redeem the hopelessly passive reader. Reading it is good for him/her. Like Baudelaire's "Fleurs du mal," what we might call Fermor's flowers of pleasure call on the doomed and the damned to wake up and know that there is something good, that something moves us on, with the help of this shower of gratification, and armed with our new understanding of language, to reassert our grip on experience. We begin to have a way. Fermor's travels are not only his own.

Obviously, one could not move directly from Fermor's gorgeous language to the restrained vocabulary of the sciences without producing a painful shock. But, even in going from the terminology of literary criticism to any semi-scientific vocabulary, one encounters obstacles that appear insuperable. While writing about Fermor, I felt entitled to draw conclusions from my reading that no one else might share, but I did not think that that fact would disqualify them. The same could not be said of a scientific thesis, which requires consensus at some stage. Even if one could find a more nearly objective, say, cognitive neurolinguistic equivalent for any of my inferences (e.g., that Fermor shows us how the destination of a sentence is created by the sentence itself), other neurolinguists would not necessarily share the appraisal of Fermor's style on which that inference would rest. As for the "moral" value that I find in Fermor's style, that lies at a still further remove from neurological analogies, though a field of neuroethics is beginning to take shape (see Tancredi, 2005; Gazzaniga, 2005; and Hauser, 2006).

In a field of great travel writers that includes Rory Stewart, Colin Thubron, and Robert Byron, all of whom traverse the same geography, Fermor is distinctive, and it is still that distinctiveness which the humanist critic engages. Despite the long shadow cast by Duchamp's urinal and by Walter Benjamin's essay on art in the age of mechanical reproduction, traditional criticism is still grounded in the study and appreciation of the individual work. Biology has certainly devoted itself to the study of individual organisms, and, for that matter, there may be only "individuals" in the universe as a whole; all our laws have flaws. On the whole, though, at least at its present stage of development, neuroaesthetics may manage to discriminate quite finely among related states of mind on the basis of their neurological representations (e.g., among various states of excitement—the excitement of metaphor, of intellectual discovery, of sex), but it does not deal with the nitty-gritty of one's excited response to a single work. Old-fashioned criticism depends on an entanglement with the individual poem, piece, or picture, for which there is no substitute. That is its laboratory, from which a generalization about art may emerge (see Chapter 1, Section 2c).

A curious feature of metaphor reflects the relation of the general to the particular. The "vehicle" of metaphor is never a particular: it is always a class or a generalization. (By using the term "vehicle,"

I follow the practice of I. A. Richards [1965, p. 96]; in the phrase "Achilles is a lion," Achilles is the tenor, and the lion is the vehicle.) Through an interesting coincidence, two years after I published this observation (1999), Sam Glucksberg (2001, pp. 37–40) arrived at a similar conclusion. How the tension between tenor and vehicle in metaphor could be represented neurologically is a question to which I am currently seeking an answer.

4. MUSIC

a. Levitin on Rachmaninoff and Haydn

In comparing neurological and humanistic views of music, it is instructive to begin with an example from Daniel Levitin's interesting work, *This is your Brain on Music* (2006), which illustrates how a trained neuroscientist who is also a trained musician is torn between the two approaches.

On one page (p. 89) the author offers the caricatural version of music criticism that neurology uninformed by taste (which is assuredly not Levitin's case) is bound to produce:

> When I hear Rachmaninoff's Piano Concerto no. 3, the hair cells in my cochlea parse the incoming sound into different frequency bands, sending electrical signals to my primary auditory cortex—area A1—telling it what frequencies are present in the signal. Additional regions in the temporal lobe, including the superior temporal sulcus and the superior temporal gyrus on both sides of the brain, help to distinguish the different timbres I'm hearing. If I want to label those timbres, the hippocampus helps to retrieve the memory of similar sounds I've heard before, and then I'll need to access my mental dictionary—which will require using structures found at the junction between the temporal, occipetal [sic], and parietal lobes. So far, these regions are the same ones, although

activated in different ways and with different populations of neurons, that I would use to process the car horn. Whole new populations of neurons will become active, however, as I attend to pitch sequences (dorsolateral prefrontal cortex, and Brodmann areas 44 and 47), rhythms (the lateral cerebellum and the cerebellar vermis), and emotion (frontal lobes, cerebellum, the amygdala, and the nucleus accumbens)—part of a network of structures involved in feelings of pleasure and reward, whether it is through eating, having sex, or listening to pleasurable music. (p. 89)

But, as if responding to a guilty impulse, on the next page Levitin gives us his own, personal response to a piece of music, one unmediated by neurological insights:

In the "Surprise Symphony" by Haydn (Symphony no. 94 in G Major, second movement, andante), the composer builds suspense by using soft violins in the main theme. The softness of the sound is soothing, but the shortness of the pizzicato accompaniment sends a gentle, contradictory message of danger, and together they give a soft sense of suspense. The main melodic idea spans barely more than half an octave, a perfect fifth. The melodic contour further suggests complacency—the melody first goes up, then down, then repeats the "up" motif. The parallelism implied by the melody, the up/down/up, gets the listener ready for another "down" part. Continuing with the soft, gentle violin notes, the maestro changes the melody by going up—just a little—but holds the rhythms constant. He rests on the fifth, a relatively stable note harmonically. Because the fifth is the highest note we've encountered so far, we expect that when the next note comes in, it will be lower— that it will begin the return home toward the root (or tonic), and "close the gap" created by the distance between the tonic and the current note—the fifth. Then, out of nowhere, Haydn sends us a loud note an octave higher, with the brash horns

and tympani carrying the sound. He has violated our expectations for melodic direction, for contour, for timbre, and for
loudness all at once. This is the "Surprise" in the "Surprise
Symphony." (pp. 90–91)

Although the above passages provide what might be considered
an adequate example of the kind of contrast that I am attempting
to illustrate, I still feel obliged to offer a sample of my own response
to a piece of music. It is not only that analyzing the differences
between Levitin's passages would be too easy, yielding only obvious results. It is also that, in order to speak with conviction, I will
have to work through some musical materials myself. I will have
to see what problems emerge when I discuss a piece of music and
then compare the results of that exercise with an understanding of
neuroaesthetics that is more complex than the one in Levitin's first
passage. Hence the following reading of Schubert's "Die Forelle"
["The Trout"].

b. Schubert's "The Trout"

The quintet, which was developed partly from the song of the same
name, is, of course, one of the great masterpieces in the entire repertoire of chamber music (Figure 12). The song's text, by Christian
Friedrich Daniel Schubart ([1783] 1968), is not as well known as the
music. What is certainly not common knowledge is that Schubert
chose to omit the last stanza, which changes the tone of the poem
completely. In the verses that were first published, the pastoral scene,
in which a trout is caught, is just a preamble to a crudely stated
moral, namely, "Girls, watch out for seducers!" The last stanza reads:
"Ye who linger by the golden spring / Of secure youth, / Just think
about that trout! / See the danger, and run! / Usually you fall / Just
by not being clever enough. / You girls, watch out for / The seducer
with his rod!— / Otherwise, you will bleed [bloom?] too late." ["Die
ihr am goldnen Quelle / Der sichern Jugend weilt, / Denkt doch an
die Forelle; / Seht ihr Gefahr, so eilt! / Meist fehlt ihr nur aus Mangel

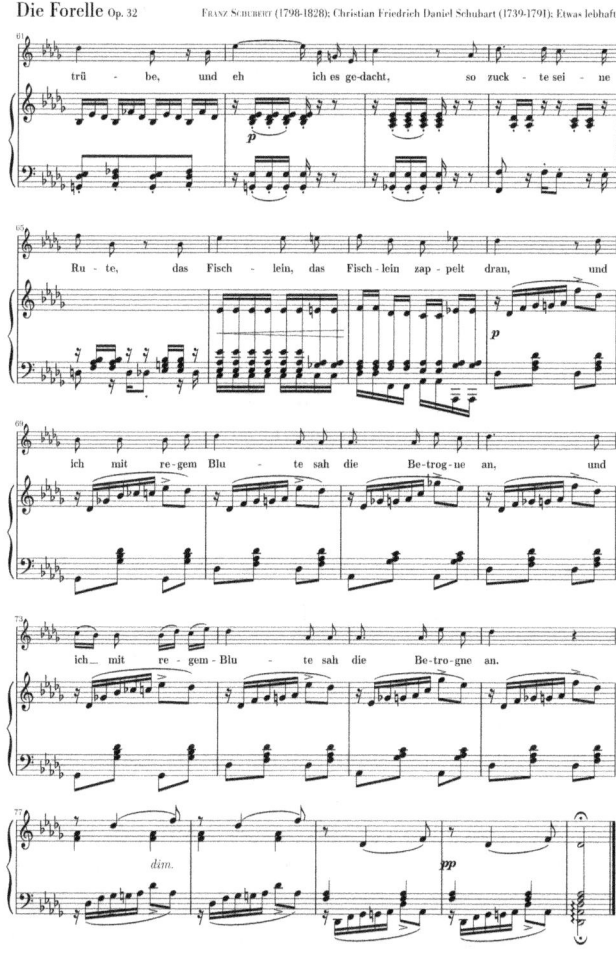

FIGURE 12. Schubert's "Die Forelle." Typeset by Maurizio Tomasi as part of the Mutopia project (http://www.MutopiaProject.org/).

/ Der Klugheit. Mädchen seht / Verführer mit der Angel!— / Sonst blutet ihr zu spat."] The last line is obscurely worded, and is perhaps simply unintelligible. (There are also traces of an alternative stanza [see Zilkens, 1997].) In any case, the whole poem is revealed to be a terrible piece of didactic doggerel, with a scarcely concealed Freudian subtext. One has to read the poem in its entirety to realize how bad it is, because in its truncated form, with Schubert's wonderful music unavoidably associated with it, it seems almost tolerable as a piece of mild pastoral with a melancholy ending. The composer certainly showed good taste in omitting the last stanza; even his transforming genius might have had trouble redeeming the whole of this poetic failure. By deleting the last eight lines, he shifted the emphasis of the text decisively. I will try to follow the effects of the music's interaction with this altered text.

The outline is simple: the speaker (presumably the poet) is watching a trout playing in a brook. The trout shoots by "in happy haste," like an arrow (line 2). It is characterized, first, as "launisch" ["tricky, sly," in line 3], but also as "munter" ["cheerful," "lively," in line 7].

As the poet proceeds with his description, we discover that there is a fisherman standing by the brook, also watching the trout, but "mit kaltem Blute" ["in cold blood" (line 11)]. The sympathetic poet is thinking that, as long as the water remains clear, the trout will not let itself be caught, but, before he knows it, the fisherman, grown impatient, muddies the water. Sure enough, before the watcher is fully aware of what is happening, the little fish is quivering at the end of the line: "und ich mit regem Blute / Sah die Betrogne [sic] an." ["And I, my blood astir, / Found myself looking upon the one deceived" (lines 23–24).] We should not make too much of gender identifications in German, but it may be of some slight interest that the gender has shifted from the neuter ("das Fischlein" ["the little fish"]) to the feminine ("die Betrogne" ["the one deceived"]).

There is no need to expatiate on the musical accompaniment, which mimics the sound of water in its fluidity, in its hesitations and in its accents. The melody is in perfect balance with the accompaniment: cheerful, steady, progressive, as of someone strolling with a lightly swinging step along a brook. We are simultaneously lulled and kept moving by the sounds. The wonderful lilting motif (mainly in the keyboard's treble part) provides a delightful momentum to the action. Yet that lovely purling motif, suggestive of a breaking wavelet, also curls back on itself, in a hooklike profile, and, at the end of the song, the potential threat in that retracted contour becomes all too apparent.

When we begin to realize that the idyll is threatened by the actions of the fisherman, a slightly anxious, rushing movement makes itself felt in the keyboard part; the melody ceases its flow and begins to fall into the accents of speech. Suddenly, we recognize that the text is no longer subordinate to the music. Speech threatens the security that music (the unmuddied brook) had provided.

The agitation increases as the action speeds up: there are a few bars of palpable urgency; the melody becomes syncopated. Up to the last moment one hopes for a rescue, if not from what the words are saying, at least from the composer, by some musical gesture that will compensate for the loss of the trout, or at least soften the blow.

The opposite is what happens. Instead of slipping into a minor key or offering some gesture expressing sorrow or sympathy, Schubert signals his intention of returning to the unruffled, resolute progress of the brook by introducing a precadential V–V7 sequence. He then lands squarely on the tonic major chord at the very moment when the victim's fate is sealed, and launches once more into the purling little keyboard arpeggios. The death of the trout is the note on which the cheerful progress of the accompaniment, briefly interrupted by the fisherman's triumph, is resumed. Now, of course, that accompaniment is felt as the indifferent cruelty of nature, rather than the supporting movement of a happy flow. Now that brook is taking us where we no longer want to go. We are the "betrogene," the betrayed.

But is it by nature that we are betrayed, or by the composer? It seems that Schubert himself is the fisherman, and we are the trout. It was, after all, his choice to cut the ground out from beneath our movement of sympathy toward the victim; it was he alone who decided not to allow us even a pause for regret, an instant of mourning. His resumption of the steady pace comes one beat too soon. It bespeaks a profound coldness, a willingness to let things go. As does the song itself. In a callous insult to the observer-poet, it overrides his dismay, forcing him to sing repeatedly "Und ich, mit regem Blute / Sah die Betrog'ne an" to the same jaunty tune that had accompanied the happy beginning of the song. It stops only when the repudiation of sympathy has been made completely clear. Then the accompaniment, too, mutters to a stop. There is no need to go on with the meaningless, mercilessly upbeat repetitions; why labor the point.

Perhaps it is wrong to make this song bear so much weight, especially since Schubert developed it into the immense masterpiece that is "The Trout" quintet. Surely, Schubert, not notorious for villainy, can be allowed his moment of emotional dissociation. And can one hold a piece of music to a moral standard? Besides, does not the greatness of the song lie precisely in its realism, in its refusal of sentimentality, in the daring move that Schubert makes by returning it prematurely to the register of impersonality, to the brook?

For me, the weight still falls on the negative side of the scale. That last note of the victim's fate, used as the first note of the song's resumption, is a betrayal of art itself as well as an outrage against ordinary human feeling. One of the things it says is, "This is just a song: I can do whatever I want in it; this isn't really a moral issue; it's just notes."

Maybe, then, it is not really a betrayal, but just an acknowledgment of music's limits. As such, this perhaps trivial example would open a window or, at least, a peephole onto the vast issue of artistic responsibility. Clearly, the artist fulfills that responsibility only by producing something that we experience as authentic, moving, and necessary. Yet, since Plato, there has always been the question of how far we should trust the work of art. Perhaps "The Trout" reminds us that we should not trust it too far.

There sometimes comes a moment, in our experience of art, when we are reminded of the medium in which it is presented, or of the form in which it is cast. At such times, our awareness of the form or medium may temporarily outweigh the effect of what is reaching us by those means. Even Constable's cloud studies are just paint. Even Mozart's operas are just notes. (On the balance between convention and independence in *La Clemenza di Tito* [*The Clemency of Titus*], also pivoting on a very brief passage, see Currie, 2002.) Even Tolstoy's novels reveal a rhetorical structure if one looks closely enough. Sometimes artists let the façade of total commitment slip, rather as Othello in the end reveals his inner monstrosity. It is just a game, we know. Let us not pretend that it is the real thing. Even as I write now, I am aware of a certain playful narcissistic pleasure at the thought that I may have discovered something, a narcissism that vitiates whatever discoveries my writing might lead me to.

Is this really a universal flaw, though, or a weakness that can be transcended? I will not try to answer that question, although I think it not without significance that I feel no sense of unwholesome artifice, and no guilt, nor any afterthought of guilt, while attempting to improvise music. Perhaps the weakness, or flaw, can become apparent only in retrospect to myself, albeit no doubt immediately to another observer. Still, perhaps such an infinitesimal example of apparently guiltless creation could be taken into consideration when dealing with the broader issue of the guilt, or guiltlessness, of art.

~~~~~

To return to the matter of comparisons, what points of comparison or contrast, if any, can be found between my hastily improvised essay on "Die Forelle," and my Chapter 4, above, on the relevance of neuroscience for the experience and understanding of music? The problem of neurological localization—either verbal, musical, or jointly musical and verbal—obviously does not arise in humanistic criticism. On the other hand, my earlier citation of the line from Shelley, "Our sweetest songs are those that tell of saddest thought" (Chapter 4, Section 6, above), may have some relevance for my interpretation of "The Trout." If that sentiment is connected to a right-hemispheric dominance for music, then the feeling that we have been cheated of our salutary experience of melancholy by Schubert's returning prematurely to a cheerful, lilting melody may reflect the disappointment of a melancholy-prone right hemisphere (or, at least, of a sympathetic "mirror neuron").

More broadly speaking, as a matter of contrast rather than comparison, the entire evaluative activity that is unavoidable when we speak of art has been, so far, largely out of bounds for neuroscience. For instance, I have expressed my opinion about the value of Schubart's poem, which, especially in its original form, I hold to be virtually beneath contempt. One probably would not find a laudatory essay devoted entirely to the text of "Die Forelle." The song, on the other hand, is an acknowledged masterpiece that turns a trivial bit of verse into a major artistic event. Nevertheless, both judgments are subjective and are not susceptible to proof. There have been enough reversals of aesthetic taste in history (for instance, in the relative rankings of Bunyan's *Pilgrim's Progress* and Addison's *Cato*) to make anyone with some experience in cultural history careful about making dogmatic assertions concerning the self-evident rightness of an appraisal. In this case, further widening the margin of subjectivity, I have even added a moral evaluation to the aesthetic one: I have actually attached moral blame to Schubert for an aesthetic choice he made at a certain point in his composition. The standards of moral right and wrong, in cases as subtle as this, are at least as far from scientific determination as is the aesthetic value or inadequacy of particular works.

What of the word-music relation, a major topic in Chapter 4, above? My argument that "Die Forelle" illustrates the ease with which music can manipulate words may be circular, but I find it hard to see what alternative interpretation there could be of what happens when the two media are brought together in this song. As long as the words and the music reinforce each other, there is no problem; they carry the same mood, though the music does so more effectively. At one point in the song, when the jerky rhythms of speech intrude, as the fisherman makes his move, we see once more the principle that music is associated with the good, language with the bad. Still, even in this passage, the music is working with the words, not against them. At the end, though, something different happens. For the first time, the mild tone of description in the text is disturbed. If there is anything at all notable in this feeble poem, it is the half-line, "Und ich, mit regem Blute" ["And I, my blood stirred up"] (line 23). Here at last we feel that something serious is being experienced; that there is something actually at stake. If this slight poem has a climax, it is to be found in these few words.

But this is, of course, *after* Schubert has withdrawn his support from the spectator, refusing him his moment of sorrow. In terms of the music, not only is it not the climactic moment in the song, it is already part of the trailing-off effect, in which the poet/plaintiff is forced to participate in the resumption of the brook's casual progress. As I have said before, the words "Und ich, mit regem Blute, Sah die Betrog'ne an" has to be sung to the same cheerful tune as the opening bars of the piece. Once again, the words are at the mercy of the music, but in this case it is the music that is turning the tables, that is betraying the language, not the language the music (cf. Chapter 4, above).

Professor James Currie proposes a much subtler formulation of the moral issue in "The Trout" than the one that I offer. Currie suggests that what is on trial in this song is neither the fisherman nor the composer, but the song's own musical style: that the target of its moral criticism is the seductive ("sly," "tricky") kitsch that characterized much Viennese music in Schubert's own day; the song both exemplifies and punishes this dishonest undertone in its own manner. This is a critique so profound that it had better be conveyed in Professor Currie's own words, and it is my hope that it will be published shortly.

All in all, then, I am inclined to say that my argument in Chapter 4 concerning the ways in which right hemisphere dominance becomes apparent in song is borne out by my analysis of the movement of Schubert's work. As usual, though, a general question remains: What is the major difference between the method by which I reach my conclusions in Chapter 4, the "Neuroscientific" chapter, and the method by which I do so in this chapter (Section 3b), where I offer an "impressionistic" or "humanistic" discussion of the word-music combination? The principal conclusion in both chapters is, perhaps predictably, the same: music outweighs text. The difference between the methods by which I reach this conclusion is the one that I discuss in Chapter 1 as well as in Section 3a, above, and it holds true for any comparison between any kind of scientific, and any kind of humanistic, investigation. With the humanistic method, one may succeed in deducing great things from small, as scientists from Newton to Einstein are reputed to have done; but one's conclusions, whatever they may be, are the byproduct, not the objective, of one's having immersed oneself in the work for its own sake (cf. Chapter 1, Section 3, above).

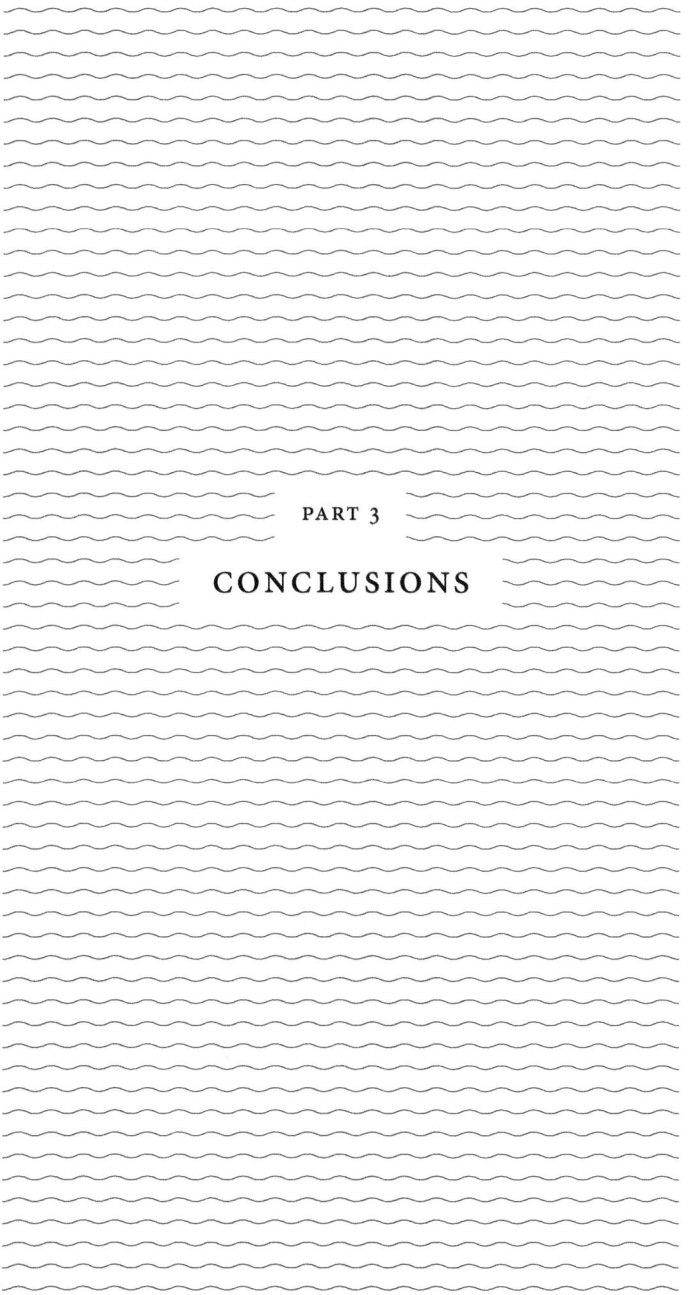

PART 3

CONCLUSIONS

6

~~~

Ideas and Values

There is no way to deal satisfactorily with the problem of values—for instance, the problem of tragedy treated as a value—within the structure of this book. Yet it is the very fact that values cannot be accommodated comfortably within this structure that defines the rest of the book. One can talk about alternative, i.e., either neurological or humanistic, approaches to the plastic arts, the musical arts, the language arts, but there are no alternative approaches to values. The domain of values lies at the outer edge of all the subsidiary processes that lead us into action; it represents the last judgment, what takes place after we have taken all the contributing factors and influences into consideration. For instance, being aware that a sniff of oxytocin may have made us more trusting than usual, or that a dose of testosterone may have made us more aggressive than usual, we still have to decide whether a certain person is a saint or a sinner, and we still have to feel—and act—accordingly. One cannot dissolve a value judgment into the influences that affect it. In the end, a judgment is made, whether by taking all the relevant influences into account or by ignoring them. Whatever neurological activity may accompany our value judgments, it is discounted at the moment of choice. Whatever the reasons why we accept or reject something, we have to end up accepting or rejecting it.

Our values and preferences may even determine what we identify as the influences on that very decision. There is an endless interplay between our choices and our understanding of the environments that affect those choices (cf. B. H. Smith, 1988). Finally, though, there is no choice. We want something. Call it Spinoza's "conatus," or Leibniz's innumerable little desires that add up to what we call behavior. (For the historical background to Leibniz's "petites perceptions," see Gross, 2006, Chapter 1.)

One form that desire takes is our desire for art. There is no easy (or, perhaps, even any difficult) explanation for that desire. But the question of why we desire art is liable to be understood instrumentally, as "What purposes do the arts serve?" Or (heaven forfend) as John Carey (2005) would have it, "What is literature good for?" And the answers might come out as "Art satisfies desire," "Art gives us a competitive advantage," "Art enhances the probabilities of reproductive success," and so forth.

2. THE PURPOSES OF ART/
THE LITERARY ABSOLUTE

Sure, music may enhance the likelihood of reproductive success. So will drinking Gatorade. This does not mean that music is Gatorade. Sure, music affects the same brain areas that light up in response to chocolate, sex, and drugs. Again, we need not rush to the conclusion that music is chocolate. To take another example, the fact that metaphor may illustrate the "appeal of the rare" (see Chapter 3, Section 3, above) does not mean that it affects us in the same way as does the sudden appearance of a stingray under foot. Art is not good for us in the same way as is Gatorade, chocolate, or the attention/startle response that saves us from the stingray.

No. The fact is, art has nothing to do with any of these things. To speak only of music. The discovery of a musical phrase that did not exist previously (and I use the word "discovery" advisedly) is a kind of "affordance" that may help us to learn not so much what art is good for, as what life is good for: namely, that it enabled us to discover that particular musical phrase. I conclude, then (shades of the 1890s), that life exists in order to make the artistic experience—and, perhaps, certain other experiences—possible, not vice versa. What we need to find is the justification for life, not just an explanation for the curious, incidental fact that it tends to perpetuate itself.

Having reaffirmed this old-fashioned, commonsensical, not even particularly aestheticist opinion—that art is good in itself, and serves no Darwinian purpose—I have been wondering about a possible hierarchy among these purposeless purposivenesses (to invoke a barbarism for which Kant is responsible). Nor do I feel particularly guilty about having abandoned, on such short notice, the quest for a utilitarian answer to the first question and skipped to a second, entirely impractical one. The current insistence that everything meet the Darwinian test seems more and more pathetic, inappropriate, and Quixotic at a time when that misconceived experiment in evolutionary biology, humanity, seems less and less viable. Either God or Darwin got it wrong this time.

So what I will think about is not, what are the arts useful for, but what kind of music or literature sets the standard; what is an essential piece of music or an essential poem; and what sets the standard for what poetry is.

In a way, this issue comes down to the ancient competition between tragedy and comedy, the *Iliad* vs. the *Odyssey*. I have noticed that, when things become too difficult in my life, only contrapuntal music will do for me, will help me. Late Beethoven is too original. One could well claim that late Beethoven is better than Bach because Beethoven breaks boundaries and channels a genius as great as Bach's into entirely new and inimitable processes. For all its daring leaps and straining harmonies, the "Agnus Dei" of Bach's "B Minor Mass," which for me is the pinnacle of Bach's expression, only stretches conventions; it does not breach or actually violate them. Yet it is best for me when I am in trouble. It holds me in, contains me while it expands me and all about it. It is essential music.

I will not attempt to make a similar comparison for art, simply because I cannot think of an artist who occupies a position similar to that of Beethoven. Mantegna? Vs. the earlier Fra Angelico? It does not do the same thing. What I do know is that I think au-

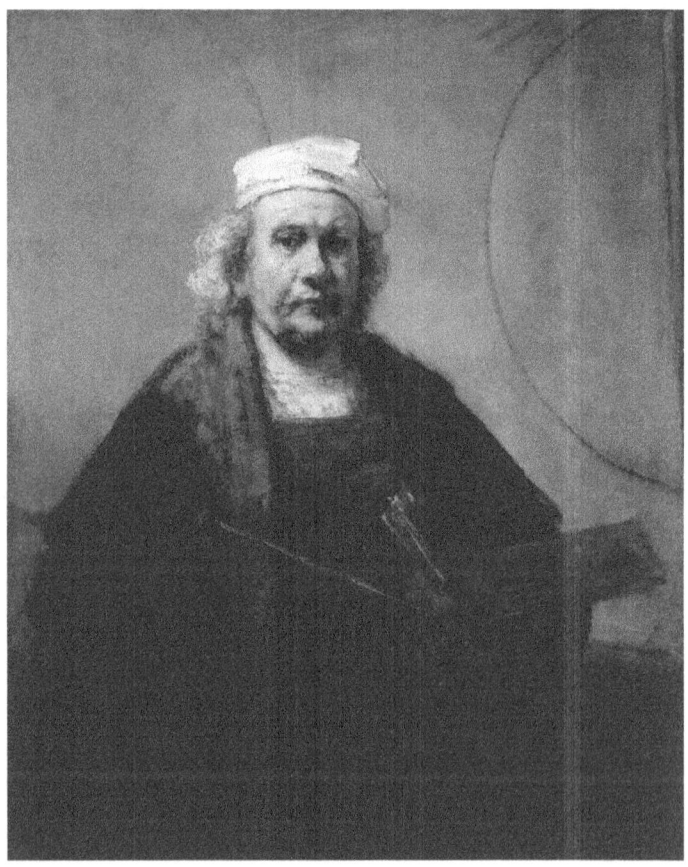

FIGURE 13. Rembrandt, Self-Portrait, *c. 1665. Oil on canvas, 114.3 × 94 cm. The work forms part of the Iveagh Bequest, at Kenwood House, London.*

tomatically of Rembrandt when I *need* a picture: the portrait of a young woman in Montreal or the late self-portrait that hangs in Kenwood House, London (Figure 13). No big surprises there.

Since I will soon be talking about tragedy, it would be fair to ask where Bach and Rembrandt, or, more generally, music and art, stand in relation to tragedy. This is a difficult question that can have only a wrong or a stupid answer if one attempts a short reply. Allowing for this difficulty, I would hazard the suggestion that music is rarely only tragic. (I think of the second movement of the Schubert Piano Sonata in A Minor as a possible exception.) Music is usually good for one, and, according to the concept of tragedy that I try to develop later, the tragic is never good for one. But the whole question of what emotions music represents, if it represents any in the usual sense, is severely vexed: for instance, a piece in a minor key is not necessarily "sadder" or more expressive of profound and sobering feelings than is a piece in a major key.

As for what counts as essential art, I am again not sure. The Rembrandt Kenwood self-portrait is for me essential as Antonello da Messina's *Christ Crowned with Thorns* or Matthias Grünewald's *Isenheim Altar* is not, although the latter are much more obviously "tragic"; perhaps here I fall back into Winckelmann's neoclassical camp. In any case, what I have been calling the essential in both music and art is more liable to be contaminated with benefit than essential literature needs to be. This probability should have been evident to me earlier, when I began my argument regarding essential music and art, from the mere fact that I spoke of *needing* the B Minor Mass' "Agnus Dei" or of *needing* a Rembrandt, as though to repair something in me. I do not think of the tragic in literature in that way. If it helps me, it helps me in spite of myself.

~~~~

What, then, of the tragic in literature? In my section on Patrick Fermor's travelogues (Chapter 5, Section 3b), I describe, at some length, the delights of reading.

What, then, of the pains of reading?

I am inclined to call Fermor's work comedy, not (according to the classical definition) because it has a happy ending, but because it takes our side, or rather, lets us take its side. We stand and move with the account and its author. We are not rejected. His account invites us to go with him; it does not repudiate us. He does not stand alone, or want us to stand alone.

The alternative is mere, sheer pain. We can fairly say that Fermor's writing is good for us. We may even be able to say, at an extreme, that a poem such as Elizabeth Bishop's "One Art" (1983, p. 178) is good for us. But pure tragedy? That is not good for us. It is neither good nor bad; such terms become irrelevant.

One may say that everything in literature except tragedy enhances fitness. Tragedy does not enhance anything. It tells us what is happening. It says what there is, but it does not assume that we can or should do anything about it. Nor does it look far beyond itself. Tragedy has no time to waste on eternity. Eternity vanishes in the urgency of the present, though it continues to hang around behind tragedy. (Perhaps it is the awareness of that fact that makes *Hamlet* into something other than a "normal" tragedy. The play questions our aspiration to the metaphysical as well as our hope for a tolerable human condition.) Tragedy has all the limitations of mortality, although it forces us to forget our mortality. It makes the present everything. And so it is.

As a more or less arbitrary sample of the tragic in literature, I can use a short poem by a scarcely known Yiddish poet; it says all that is needed to make my point. The poem's author, Moishe Shaffir (1940), suffered a particularly crushing series of disasters in his life. A child with Down's syndrome, whom Shaffir raised at home, along with his own gradual descent into blindness, left trails of sorrow in his poems. His last plea begins,

> I reach to you with these my limbs in prayer;
> I fall away in dust before your feet— (lines 1–2)

Then he begs God to put an end to his poetic impulse once and for all—never again to let the dew of poetry fall upon "The downslopes of my soul" (line 6)—and to do away with the visions in his dreams that attack him with burning spears; they "tear the tender strands of sleep" (line 10):

> The night is full of autumn howling;
> Day drips as from a naked blade.

Strike my eyes, grown dark, with blindness:
Strike, strike them, father, with your gentle hand.
   (lines 15–16)

(In the original, the last two lines read, "Shlog mayn krank-gevoren oyg mit blindkayt / Shlog es mit dayn foterlikher hant." I do not pretend to have translated Shaffir's poem adequately.)

Initially, I intended this poem to be the culminating example in an essay on the theme of illness in Yiddish poetry, and, when presented as a freestanding poem, it cannot retain the demonstrative force that it held in that context. Nevertheless, it says what is necessary. There is nowhere to go beyond this poem, nothing further to appeal to. It is simply a threshold. This is the "literary absolute" (cf. Lacoue-Labarthe and Nancy, 1978). This is what poetry is for; this is what poetry, when it is fully poetry, does.

In the context of my general project—comparing the neurological with the "straight" humanistic approach to the arts—this section represents the antithesis of a strictly scientific method, is its extreme and defining alternative. It is not concerned with the "how" by which the artistic event is initiated or appreciated, or with the "why" that might explain its emergence in evolutionary or survivalist terms. It assumes the artistic event to be a good in itself, or, since "good" may imply some pleasure or advantage (which art does not always provide), an end in itself. Sometimes, just an ending.

3. THE PHILOSOPHICAL SUBSTRATE

It is not clear to me where tragedy falls in relation to philosophy; in any case, I have allotted tragedy a separate place. Even if it could be assigned a position within philosophy, it is sufficiently distinct from the other philosophical problems I reconsider in this section that it seems to exist in a separate domain.

Having demonstrated, to the best of my ability, two ways of thinking about the imagination in the various arts and in several subfields of aesthetics (and, coincidentally, ethics) about the imagination, I feel obliged to review some of the general philosophical problems that subtend such an inquiry; after all, some would say that neuroscience itself is just an anthropologizing of philosophy. These general problems are involved in everything that I have been discussing, but they remain undefined. Perhaps it would be best to leave them that way, to let these shadowy sleeping dogs lie, since they belong to the family of problems that even real philosophers have never come close to settling. On the other hand, it would feel cowardly to leave them lurking in the background, without at least letting them come out of the dark and give themselves a name, even if it be only a non-name such as "the mind-body problem" or "the nature of perception."

Of course, all these matters have been addressed by professional philosophers, from Aristotle, who mapped the whole scene, through the moderns—for example, Hilary Putnam, on thought and language, or Gerald Edelman and Nicholas Humphrey, on qualia—to the post-connectionists, who examine the core problem of representation. There are now even neurotheologians, such as Richard Strassman, Michael Persinger, and Andrew Newberg, who interpret our religious experiences neurologically. My only excuse for inflicting my thoughts about these daunting topics on the reader is that my project would seem radically incomplete if it did not include some account of my own struggle with the fundamental issues. I certainly do not pretend to have solved the ultimate problems with the certainty of a Daniel Dennett (2005) or of a Bennett and Hacker (2003).

4. DIDO AND AENEAS

Whatever understanding, whether simplified or highly sophisticated, neuroaesthetics adds to traditional discourses about the arts can only be beneficial; it certainly can never render aesthetics obsolete. Humanists may fear that their embattled island of subjectivity will vanish amid the rising tides of science, but, in the end, scientists do not forget that the arrows on their meters and the shadows on their screens pertain to their own living, or dying, consciousnesses—to their own subjectivities. The arrows may be objects, but what they point to is not. We know that it is not the rod, the cone, or even the retina that sees; it is we who see those arrows and those shadows. And it is at us that those arrows point.

I always come back to Purcell's *Dido and Aeneas*, specifically to "Dido's Lament," which has haunted me all my life: "When I am laid in earth / May my wrongs create / No trouble, no trouble in thy breast; / Remember me, but ah! forget my fate" ([1689] n.d., Act 3, last aria). When I listen to that song, I like to think that every neuroscientist would share my opinion about it in relation to the scientific enterprise. Singing those words, hearing the music to them, and understanding and feeling what they say—experiencing them, in a word—I cannot help but believe that any attempt to translate that meaning and that experience into neuroscientific terms would be a desecration of the music and a violation of our human spirit.

What does "Remember me, but forget my fate" have to do with neurology? Dido knows that she is about to be laid in earth, in that dumb medium that is unable to sing. This she laments. In our stupidity we would try to reduce her song itself to earth, to a physical "substrate," to some sort of object that we can take hold of. If so, we have missed the entire meaning of her song; and we will never remember Dido. Ourselves suffering from some sort of neurological deficit, we cannot grasp, cannot apprehend, what is being offered to us: that her song remains when the earth of which her body is made is gone and forgotten. Earth cannot sing her song for her; a neuron cannot sing her song for her; as we all should know, only a human being can.

It does not matter how deeply we analyze functions, finding a neuron for this and for that: a Bill Clinton neuron, a grandmother neuron, a mirror neuron, a sympathy neuron, a music-seeking neuron. None of these will tell us how "Dido's Lament" will sound to us, or how we should experience it. It alone can tell us what to feel.

5. SEQUENCE

[W]e now lack any reason *beyond mere correlation* to think
that a neural system could be the substrate of (the sufficient
basis for) an experience. (Noë, 2004, p. 247)

I have said that when we observe animal behavior, or when we remark on anomalous human behavior, we say without hesitation that
that behavior has been *caused* by certain neurological processes.
Intervention in any well-known neurological circuit makes this
connection abundantly clear. Ribot (1885) points out that the experimental method itself rests on a causal hypothesis. Not only is
every mental state invariably associated with a state of the nerves
["tout état psychique est invariablement associé à un état nerveux"
(p. ix)], but "Since every experimental method rests definitively on
the principle of causality" ["Toute méthode expérimentale repo-
sant en définitive sur le principe de causalité" (p. xix)], physiological psychology has nothing but causes and effects to work with. If a
bull suddenly becomes afraid of a mouse, it is not because the bull
has undergone some spiritual transformation that makes it realize
the mysterious power inherent in tiny creatures, like Turnus, in the
last scene of the *Aeneid*, battered by the Fury in the form of a tiny
screech-owl; no, it is because somebody has thrown a switch and
sent a current through a particular batch of neurons in its amygdala.
So what is the issue?

Perhaps because of the ease with which such demonstrations
can now be carried out, the Cartesian mind-body division has
become particularly unfashionable; for instance, Nicholas Humphrey (2006, pp. 124–134) argues in all seriousness that dualism, far
from being a respectable if disputed intellectual position, is merely
a beneficial illusion designed by nature to enhance our chances of
survival. Still, this illusion may be worth revisiting, even if only to
see how it could once have been taken seriously. After all, although
systematic neurological science was rudimentary, even in the sev-
enteenth century the effect on behavior of various substances and
various interventions, such as alcohol or a knock on the head, must

have been apparent to everyone. (The story that the seventeenth-century philosopher Malebranche became a genius after a fall from a horse is in itself significant [see Massey, 1954, p. 337, fn. 57; p. 336, on Franz Joseph Gall on similar phenomena; and p. 337, fn. 58, on André Grétry's trepanning]. George Eliot [1885, p. 332] speaks of disease as releasing mental powers. Draaisma, [2004, p. 91,] says that savant skills can sometimes emerge as a result of brain trauma, and Sacks [2007, e.g., p. 157] gives several examples of such skills developing after a cranial insult.) Avishai Margalit (2005) gives a sketch of Cartesian dualism in its Spinozistic version; he goes so far as to say that "Spinoza's insistence on making the categorical difference between the two kinds of explanations firm is exactly what is needed in current discussions on the mind-brain relation" (p. 50). Margalit argues, for instance, that a thought in itself cannot be said to have a physical consequence (p. 50). (For a phenomenological approach to this fundamental issue, see Merleau-Ponty, 1969, pp. 160–161, and, for a recent defense of Cartesian dualism, see Robinson, 2007, p. 192.)

I will approach the problem from the opposite direction: can a physical event in the body have a mental consequence? This question can be reformulated in terms of sequence. Does a neural event *cause* me to project a metaphor, or to want to write a better sentence? Is a neural event *causing* me to have my present thought? If so, it must precede the thought. It is, of course, well known that a neural event does precede, temporally precede, every conscious intention; we even imagine, or *dream* everything we do before we do it (cf. Aristotle, 1941, pp. 597–599; J. H. Jackson, [1874] 1958, vol. 2, p. 201), and our bodies know what we are going to do before we do. In case we needed tangible proof that Aristotle was right, we now know that a monkey can move an object merely by thinking about what it wants to do, not even bothering to follow the thought by a movement of its arm (Knezovich, 2006). There is also evidence for a half-second delay in perception itself (Libet, 2004, pp. 45, 75–78). We are unconsciously retrofitted, so to speak, so that we are unaware that our consciousness is always trailing our nervous system.

From another point of view, though, this apparent sequence is of no significance. What does it mean to say that our nervous system had our thought before we did? Since a conscious thought is defined by the very fact of our being conscious of it, nothing can have preceded it. Our nervous system cannot be conscious on our behalf; no one, and nothing, can be conscious for us. *We* may be more, or less, conscious of something; no object, no matter how intricate, can supply that function. This is equally true no matter how many shades and gradations of consciousness we may identify: they all still have to be conscious (Dennett, 2005 to the contrary notwithstanding). Speaking a sentence is as good an example of consciousness in action as any.

Dreams present a related set of problems. Like our thoughts, our dreams are endogenous. How can a neural event precede an endogenous experience? Such a sequence would involve us in an infinite regress. And even if a dream were to follow a neural event, what would that prove? Either the neural event *is* the dream, or it is not. What does the neural event *affect* in order to produce the dream? Another neural event? If the neural event is indistinguishable and inseparable from the dream, how can it precede the dream?

What precedes a dream? Say, a stomachache, sexual desire, anxiety? But the physical event at the cellular level and the dream are not two separate and distinct occurrences. There is not first a stomachache and then a dream: they are the same; they occupy the same space. They arise together. Even if one set of cells fires for the stomachache and sets off a reaction in another set of cells to produce the dream, at least the second firing must be simultaneous with the dream. Otherwise, it would have to act on still another physical entity in order to produce the dream, and so we are sucked into another infinite regress—or confronted with a homunculus, with whom no one seems to be happy.

The dream starts only when the dream itself starts; there is no dream before the dream. A cell cannot be dreaming for us, any more than a cell could have been thinking for us before we had a thought. From the point of view of the dream, only those neurological processes that are absolutely simultaneous with the dream can have any possible relevance to the dream. And, in that case, without temporal sequence, where is the "causality"? For without sequence there can be no causality.

Endogenous experiences generally present this difficulty. It is easy to see how an external event would precede our response to it, but the role of delay (the delay required by causality) in a self-generated experience remains obscure. The conundrum becomes even more baffling if we consider the possibility that, at least to some degree, all experiences are self-generated. It is undoubtedly foolhardy for someone who is not a quantum physicist even to quote someone who offers a perspective on a subject as profound as this, but I cannot help noticing some remarks by Lucien Hardy that bear at least a metaphoric resemblance to what I have been talking about. Following Roger Penrose, Hardy claims, "The way quantum gravity handles information may indeed be like the thought processes in the brain" (quoted in Brooks, 2007, p. 32). The point is that in the theory of quantum gravity the idea of sequence is not operative, and causal structure is indefinite (Hardy, 2007; Daniel Robinson [2007, p. 183] also remarks on the uncertain status of causality in recent scientific thought; cf. Stapp, 2007).

There is another problem. We can see how an action—say, the movement of a finger—could be brought about by many different neuromechanical processes. It is more difficult to understand how qualia—subjective experiences—could be produced by a variety of means. (Here the work of Paul Bach-y-Rita, Tyler, and Kaczmarek [2003], on the use of alternative routes to sensation, has to be taken into consideration.) If we grant that something as subjective and specific as a work of art can be produced only by a single set of neural circumstances, then behaviorists may be authorized to try to trace a particular work of art (as well as a dream) to its neurological "beginnings," to the very specific cells that gave rise to it. But if something equally subjective—say, a sense of humor—can be traced to *different* cells in *different* systems in *different* organisms (e.g., a crow, or an octopus), what happens to the subjectivity-cell linkage? A similar question might be asked about the brain. Besides the fact that there is a constant turnover of brain cells, the brain's neuroplasticity allows it to recruit other cellular arrays to perform unaccustomed functions (Hawkins and Blakeslee, 2004, pp. 54–55, 61–62; Kenneally, 2006, p. 41; Begley, 2007; and Doidge, 2007); while performing in their new roles, these other cells do not visibly change at the histological-anatomical level. But if a bird brain, which has a structure completely different from that of a human brain, can produce dreams, and qualia, and subtle experiences at least superficially indistinguishable from those of a human being, what happens to the supposedly rigid link between highly specific neurophysiological processes and/or machinery, and the experiences (in this case, aesthetic) that result from them?

I put this question to Susan Udin, Professor of Physiology at The State University of New York at Buffalo. Professor Udin responded with two suggestions. First, concerning neuroplasticity: *every neuron has the potential to do a great many different things, and that potential may be released or strengthened under appropriate circumstances.* (I myself would think that such adaptability would entail some reconfiguration of the brain as a whole, since pathways would also have to be altered. It also raises the whole complex question of cross-modal sensation, as in synaesthesia.) Second, concerning apparently similar experiences of qualia across species: *the brains of vertebrates are basically much the same.* However, Professor Udin acknowledged that it would be a challenge to account for, say, anger, jealousy, or, particularly, a sense of humor in an octopus; though, even here, Seth Grant and his colleagues have shown that complexity is a function not only of brain size or of the number of neurons, but of the molecular composition of synapses ("Origins of the brain," 2008). We will not go into the "intelligence" of slime-molds.

But, clearly, it is not satisfactory to say, simply, that a neuron is a neuron and a brain is a brain (Fox, 2007). One begins to get the impression that there is a sort of shell game going on, in which one pursues an elusive localization through many parts of brains, many kinds of brains, many patches of tissue, and many electrochemical processes. The assumption that one can establish a unique link between a certain kind of experience and a specific neural event in a particular place in a particular kind of brain begins to seem porous. (As for wider issues that philosophers from Averroes to George Mead have raised from time to time—for example, the implausibility of assigning a geometrical locus to consciousness itself—such indiscreet questions are generally avoided during discussions of neurology [cf. Melser, 2004, and Rockwell, 2005].)

But in the largest sense, the problem with which we are involved here is, how can we "get behind" a thought—any thought—in order to find its putative neurological underpinnings? Or, to put it differently, how can neurology ever catch up with a thought? You have to have the thought before neurology (another thought) can catch up with it. One might as well say that the thought produces the neural sequence with which it is associated (Margalit's [2005] non sequitur) as the other way around. I have to have thought the thought in order for the neural processes that subtend the rest of my sentence to have been set in motion. Neurons cannot see ahead of themselves, but ideas can; even identifying a class of "prospective" neurons would not alter this fact. Leibniz's distinction between perception and apperception may also have some bearing on this question. The neurology of a thought presents a different class of problems from the one that the neurology of a perception presents, though sequence is not a negligible issue even in the latter case.

The problem with falling back on a rule of neurological "concomitance" (or, if we prefer the more technical term, "supervenience" [Rockwell, 2005, p. 70]) to replace a weakened rule of "causality" is that the point of our even having remarked on the neural events in the first place becomes unclear. Was it just to supply a fig leaf for an unsatisfactory theory? It is surely foolish to deny the relevance of our nervous systems for our experience, but to establish conclusively the nature of the positive links between them remains, in many areas, only an ideal. Some neuroscientists, among them Jason Warren (in a May 2007 personal communication), are convinced that a completely different, and as yet unforeseen, paradigm

178   will have to emerge if progress is to be made in understanding the relation between the mind and the nervous system.

The issue arises even in the domain of simple arithmetic. We have neurons that concern themselves with reasoning, and no doubt we could not reason without them. For the most part, we can assume that our reasoning neurons come up with the result 2 + 2 = 4. But what if something were to go wrong with them, and they were to produce the result 2 + 2 = 3? (Changeux [2004, p. 149] provides an actual example of this phenomenon. My daughter, Rachel, has reminded me that a version of this issue also arises in Wittgenstein [1968].) Who could tell them that they were wrong? If the result is simply the product of a physical process, and if there is no higher authority to pass judgment on it, we remain at the mercy of unreason. It would not help to say that someone else, whose reasoning cells are intact, could set the poor fool straight; for that unfortunate person, the equation 2 + 2 = 3 would be an unavoidable conclusion, no matter what anyone else said to him/her. If there is no reason outside cells, there is no reason.

6. THE NEURAL STATUS OF THEORY/ HYPOTHETICAL ENTITIES

The question regarding the neural status of arithmetic leads naturally to the broader problem regarding the neural status of ideas, including the neural status of hypotheses, hypothetical entities, counterfactuals, and other such constructs that have an ambiguous existence (cf. Goldman, 2006, p. 261, 268–269). Even if "the counterfactual imagination is rational," as Ruth Byrne (2005, p. 13) maintains, it is hard to tell what the foundation for its creations could be. Of course, neurons cannot be held responsible for the rightness or wrongness of a hypothesis, nor for the existence or nonexistence of a hypothetical entity, whether of something like Shelley's "Intellectual Beauty" in poetry, or of Lisa Randall's "Fifth Dimension" in string theory. The neural status of a hypothetical

entity is no different from the neural status of any other element
of language. Both the nature of its existence and its rightness or
wrongness are judged after its formulation or utterance, in the real
world. Its reality or lack of reality, its rightness or wrongness, are
matters that have to do with the world, not with neurons.

In spite of these fairly obvious facts, when we remember that
hypotheses exist mainly in order to be proved wrong, sooner or
later, there is something odd about assuming the existence of a
clear-cut neurological underpinning for such volatile creations,
or for demonstrably wrong conclusions. (After all, even the best of
theories has only provisional contact with "reality" [cf. Rockwell,
2005, p. 175].) We tend to think that neural processes, being actual
physico-chemical-electrical things, should be linked to something
true in the mind or to something real in the world.

Accepting the assumption that some neural event accompanies
every mental event, one may still encounter classes of mental events
that make one wonder how that partnership can work. For instance,
what could be the neural status of a concept such as intentionality
(Lloyd, 2004, p. 257), especially if we are expected to distinguish it
from the neural status of an intentional act? Again, if you posit a
neurological equivalent for twoness, you must still have *invented*
twoness in order then to discover it, whether at the physical or at
the neurological level. What, then, is the neurological status of the
invention of twoness?

What is the neurology of meaning? Or of the concept of meaning? Does meaning even trickle down to the level of the neuron?
Arnold Modell (2003, p. xii) says that "meaning cannot be 'represented' by a formal symbolic code"—that is, even assuming that
the brain possesses such equipment. (For the typical, casual use
of the term "code" in describing mental activity, see Tramo et al.,
2005. Daniel Robinson [2007, p. 189] remarks on the misleading
use of "code.") It is also far from clear what "representation" would
mean under such circumstances. As Fred Keijzer (2001, p. 240) remarks, "The presence of an inner state bearing information about
an external state is not sufficient to conceptualize this inner state

as a representation." Can the content of a concept be a neurological fact in the same way as, say, a pain can be a neurological fact? (Levitin [2006, p. 160] appears to believe that it can, as represented in a parallel distributed processing model titled "Minerva.") Like Modell, I wonder how meaning can be represented neurologically: it exists in a space of potentiality. It also always contains an element of relation; it is never in one mind only. (Rockwell [2005] is eloquent on this theme.) Besides, what kind of physical equivalent can there be to something that we invent, rather than to something that is simply *there*? Is the neurological process just a registration of the words that we produce, not of their meanings? What if there is no such thing as Shelley's "Intellectual Beauty" after all, or what if we cannot define it? If it does not exist, or if all we have is a half-dozen cloudy definitions of it, do we still have a neurological "representation" of it? And even if we could imagine such a process of inscription, could it get across the barrier, in either direction, between mind (thought) and body (neuron)? Do we perhaps somehow instruct our neurons to do our thinking for us?

It seems that the brain generates some words or other symbols, but that something else (another neuron?) must decide whether that process has been an empty exercise, a merely mechanical occurrence, or whether the words have some relation to actual experience, whether they are not mere babble and refer to something. We have to become conscious of our own words in order to decide if they mean anything. But if that meaningfulness is embedded in the original generation of the words, then something more than a mechanical event must have taken place, although we may not be able to hold our neurons responsible for the emptiness or the fullness of their acts.

The annoying problem of sequence arises here again: Which came first, the thought or the neural event? The physiologically inclined might point out that damaged neurons will produce damaged thoughts; the counterargument might be that someone (not just a neuron), as in the case of 2 + 2 = 4, has to make that determination, and so on in a regress that has to end up in a consciousness (not a neuron) that has set up the criteria for the rightness or the wrongness of the result.

The existential status of a theory is hard enough to define in philosophy; it is at least as difficult to define its neurological correlate. To begin with, there is the palpable impossibility of finding your way back to a physical analogue for an impalpable idea. Any attempt to provide neurological analogies for, say, the kind of meditation in which I am engaging right now would seem alien to the spirit of philosophical thinking. The itinerary that I am following at this point is entirely different from the itinerary by which one might reach the opinion that songs are processed primarily in the right hemisphere (see Chapter 4, above). Inspecting ideas is different from inspecting behavior. Likewise, talking about the structure of ideas or about intellectual processes is different from talking about neurology, and it is hard to get from one to the other. How can you talk about the neurology of an essence? Or of *King Lear*?

I have said that meaning resides in a space of potentiality. The unforeseen is what defines us. This is most obviously true in art. Let us call art an emergent property of the human being, to use a term in common use if not therefore particularly clear. Works of art create meanings that are simply not contained in their elements but emerge for the first time from the collision of those elements. It is hard, once more, to envisage a neural equivalent for something that is not there in the components with which one began. An utterance may have emergent properties, that is, it may have properties that are not foreseeable from the words and phrases that enter into the utterance, but that emerge from their concatenation. Again, Jorge Luis Borges' (1962) remark in "The Circular Ruins" that modeling the matter of dreams is more difficult than "coining the faceless wind" illustrates my point. This phrase is a good example of the unpredictable production of meaning. It is what *does not* make sense that makes sense, what escapes plausibility that creates plausibility—i.e., new "meaning." Humpty Dumpty to the contrary notwithstanding, at some important points in style, as in the Borges quotation, language *is* the master. Some philosophers of language argue that the creation of new meaning always depends on a necessary error—on a catachresis.

It may be that more sophisticated interpretations of brain function will clear up some misguided notions or clumsy hypotheses that have created the obstacles to understanding outlined above; perhaps connectionism or parallel processing models will have more success. I do not know whether the attempt by Artificial Intelligence theorists such as Robert Port and Timothy van Gelder (1995), Dan Lloyd (2004), and James Elkins (2008) to get rid of naïve notions of representation will clear the way for a better paradigm. One of the problems with the dynamical and the connectionist strategies is that they leave one with the impression that the only real brain is a virtual brain or a mathematical model. In any case, the dominant assumption that mental events simply are equivalent to circumscribed neural events leads, especially in the domain of aesthetics, to perplexities that cry out for a different approach.

7. NEUROAESTHETICS AND ITS DISCONTENTS

It is tempting to end my work with a section titled "Neuroaesthetics and its Discontents." All that would mean, though, is an acknowledgment that I had gone through the usual cycle that one does go through when one encounters a new paradigm and works out its implications; whether it be psychoanalysis, structuralism, deconstruction, postmodernism, or a hundred others, one eventually realizes that it does not solve as many problems as one had hoped it might. So, too, the familiar lineaments and limitations of old-fashioned aesthetics begin to show through the garments of this new discipline.

Take, for instance, the now famous "mirror neurons," together with recent studies of the trust that develops in long-distance social transactions. Empathy was already a major theme in the eighteenth-century school of sentiment, for the philosopher of ethics Francis Hutcheson, the skeptical philosopher David Hume, the novelist Lawrence Sterne, and the moral philosopher Adam Smith. In the twentieth century, beginning with the great 1907 essay "Das Wissen von fremden Ichen" ["The Knowledge of other Selves"] by the major theorist of empathy, Theodor Lipps, the role of spontaneous imitation in social interactions attracted more and more attention. It seems that we identify naturally with other people's emotions (Iacobini, 2008, p. 124, and Decety et al., 2008). It should follow that, to go back once more to Shelley in the "Defence of Poetry" ([1820] 1965, p. 41), the imagination, which enables us to feel what others feel, strengthens morality "as exercise strengthens a limb."

This automatic response may be found in dolphins, but is it true of human beings? (On the difference between controlled and automatic empathy, see Goldman, 2006, pp. 209, 220.) I am always reminded of Proust's ([1913–1927] 1966–1969, vol. 1, pp. 121–124) elderly housekeeper weeping over her reading while the servant girl is weeping over the asparagus to which she is allergic, and which the sentimental housekeeper is forcing her to clean. Maybe, as Leo Bersani (1981) concludes, imitation, the exercise of the imagination, nourishes our sadistic rather than our benevolent proclivities. And so our scientific discoveries just lead us back to Percy Bysshe Shelley in one corner vs. the Marquis de Sade in the other, all over again; or, to recall an earlier version of the same debate, should we subsidize the theater, or must we ban it? Perhaps the discovery of mirror neurons casts this problem in a new perspective; perhaps it does not.

I will recapitulate what I have said in my Introduction and at various critical points throughout this book. The certainty that right and wrong, good and bad, are decisions that cannot be ceded to any material object, no matter how intricately constructed, is the guarantor of the entire philosophical project. Surely the mind surrounds, rather than is surrounded by, even neurology. Do we need to locate the neurological equivalent for our desire to live, to experience, to understand (even though it may be fairly clear what the neurological equivalent of our *lacking* that desire might be)? That would be the only reason to look for a neurological equivalent for poetry, for introspection, for thought.

A scientific approach to the arts could imply that neurology

184   has something to tell us about the arts that we need to know, that neither artists nor audiences can do their job properly without the information that the sciences can provide. Such a belief would be both pernicious and wrong. Artists may wish to take account of scientific information, but to require that they do so would be an injurious trespass. In fact, scientific research in the arts is liable to foster the illusion that we can get a handle on aesthetics (or philosophy), that we can figure out what makes art art, even that we may somehow be able to bridge the divide between the mental and the material world. But if we start from the "higher" level (and I use the word advisedly)—if we start from within the aesthetic— the "lower" things will never catch up with us; there is an eternal asymptote, and they never get any closer.

If we begin from the other end, however, if we ask, "what are the physical sources of inspiration?," the battle is already lost; art is past, and, if we still try to argue for the arts, it will be from a position of defeat. In order to think about the experience of inspiration at all, we have to start from *within* the experience, which, ipso facto, surrounds any neurophysical account of "sources," or any other account. The work of art itself, and the experience of producing it, have to set the terms of the discussion. If all those lit-up patches on the brain-imaging screen add up to the third movement of the Haydn trumpet concerto, then, bravo!, as long as that is the criterion and the result.

8. TWO TRUTHS

Let's be honest about it: there are two kinds of truth. Two kinds of truth: and one is not meant to be measured by the other. (For a diatribe against this view, see Swirski, 2007, pp. 27–32.) The humanistic truth neither desires to be, nor can be, measured by scientific criteria, and the scientific truth cannot be measured by humanistic criteria, nor is there any reason why it would wish to be.

Even the kind of esteem appropriate for prescient scientific thinkers such as Leibniz or Lipps (see Chapter 1, Section 3, above), not to mention theorists of more ambiguous standing such as Ptolemy, Lamarck, or perhaps even Freud, is alien to the aspirations of the humanist, in the sense in which I have been using that word. For it is not merely that, in my commentary on metaphor and on music in Part 2, "The Imagination, Plain," I (as a representative humanist) was not thinking about the possibility that my ideas might eventually be approached in some scientific or neuroscientific way; it was that any such suggestion would have been positively unwelcome. The intruder would have been politely told to keep his/her blundering hands out of my business. Immersion in the phenomenology of an experience, even for the purpose of thinking about it, rules out the distance implied by a scientific perspective, with its pragmatic, experimental, or mathematical criteria of value.

Unlike the theorists mentioned above, the humanist would not *want* to be "vindicated" by science; and, by the same token, the humanist is not afraid of being proven wrong by science. When Paul Valéry said, "Personne ne médite" ["No one can meditate"], and even when I (1977) said that metaphor provides access to essence, neither one of us was hoping for support from a scientific psychology. It may seem dogmatic, arbitrary, and presumptuous, but, in a genuinely modest sense, we are a law unto ourselves, not in some mystical or religious way, but by our own criterion: the truth of our immediate response to the object. That object may be a poem, a landscape, or simply a human experience. What it brings out of us, of thought or of feeling, what it extorts from us. Any generalizations arrived at on such occasions are a byproduct of the descriptive process; they are not a goal in themselves, and, as I say in my Introduction and in Chapter 5, Section 2, above, they always come with their surrounding atmosphere and a telltale piece of their matrix attached to them. They are not meant to be contemplated naked, stripped of their origins. They are meant to provoke thought, not to settle a question. And there is no further horizon, nothing more that we are aiming for. Our "intentionality," or perhaps our condition of being, is sealed in our response.

9. JOHN KEATS TO THE RESCUE

The issue of immersion in experience vs. the view from without, of physicality vs. abstraction, of the phenomenal vs. the noumenal, has been taken up again recently by Noel Jackson (2008) in the context of his chapter on the poet John Keats. For Jackson, the particular greatness of Keats lies in his ability to confront this dilemma, of the "heard melodies" and "those songs unheard" ("Ode on a Grecian Urn" [1820] 1959, line 11) unflinchingly, and to make it the key domain of his poetry. In fact, Jackson argues, Keats succeeded in making it into the key domain of poetry as such: "the very site of literary value" and "the locus for a new kind of literary authority" (p. 184).

The fact that a new kind of authority can be felt emanating from the poetry of the famously modest Keats ("Here lieth One whose name was writ in water" [Shelley, "Fragment on Keats" (1839) 1983, line 1]) represents an important insight. I would modify it only by going one step further: I believe that, at times, Keats not only confronts the paradox of sensation vs. abstraction, but that he actually succeeds in dominating the issue. (In such a case, one cannot "solve the problem.") The key seems to be the vantage point that he adopts from which to reason, rather than the process of his reasoning itself. I will have to approach this argument indirectly, but I hope that my point will eventually become clear.

I have recently come to think that Keats is indeed the figure who holds the key to many of the issues raised in this book, in particular to the central one: the apparent conflict between the scientific and the subjective, or humanistic, approaches to aesthetics. To begin with, as Keats emphasizes repeatedly, there is not only an opposition, but also an interaction, between sensation and thought: neither can be conceived of independently of the other. In a letter dated November 22, 1817, Keats declares, "The imagination may be compared to Adam's dream—he awoke and found it truth." This is similar to what I have been saying: it is only what we first imagine that we will afterward know; thought precedes knowledge. (To be

sure, this is a Kantian sort of formulation, but not to be dismissed
on that account.) Besides, as Keats says, "I have never yet been able
to perceive how anything can be known for truth by consequitive
reasoning. . . . Can it be that even the greatest Philosopher ever arrived at his goal without putting aside numerous objections" (November 22, 1817, letter). (In a similar spirit, Henry Adams, in *The
Education of Henry Adams* [(1907) 1931, p. 457], later remarked that
an insistence on finding the truth leads inevitably to the falsification
of facts.) The reasoning process itself contains flaws, or it is subject
to distortions and interference, and we keep encountering things
that are beyond the capacities of reason to deal with. As Philo once
observed (see Massey, 1977, p. 291), a fog of unreason issues from
reason itself, while experience (what Keats called the "Life of Sensations," [November 22, 1817, letter]) surrounds thought and makes
a mockery of it. Yet all of this is not necessarily cause for despair.
Life is an endless negotiation, or rather, an endless reciprocal affair between the mind and the sensations, and is to be embraced
as such. In this passage, Keats emphasizes the provisional nature
of thought.

But what would ideas—hard, practical, scientific ideas, even neuroscientific ideas—look like from within Keats' "Life of Sensations"?
As it happens, we are in a position to give Keats' own answer.

Keats' "Ode to Psyche" provides the proverbial field day for
historians of neuroscience, and the poem has been done ample
justice by Alan Richardson (2001, pp. 124–128) from that point of
view. Keats laments that Psyche (of Apuleius' "Cupid and Psyche"
story; her name means both "soul" and "butterfly") has not been
given sufficient respect in the Pantheon, for her divinity has been
conferred on her only recently, and grudgingly. It will be remembered that her love affair with Cupid depends on darkness; when
she throws light (and, in the excitement of desire, hot oil!) on Cupid,
he vanishes. Only after she undergoes a long series of terrible trials
is he restored to her. Keats wishes to confer on her the full dignity
that is her due. In an extraordinary conceptual move, he imagines
her as the goddess of neuroscience.

For Keats, the turn toward neuroscience is not the acceptance of spiritual failure. Keats had medical training and, as Richardson (2001) has shown, was fully aware of the most advanced neurological thinking of his day. For him, it is the other gods, the "faint Olympians," (line 42) who should be dismissed. Psyche is the new great goddess; her knowledge makes her a twin to Keats' god Apollo, who says, "Knowledge enormous makes a God of me" (Keats, "Hyperion: A Fragment," [1820] 1959, Book 3, line 13). Psyche will be Keats' *goddess* of knowledge, the goddess of the most intimate and the most significant knowledge: the knowledge of the human mind, *in its material manifestation.* For this reason, he builds her temple in his mind. It is a real nervous system; it will stand in a forest not of pines, but of dendrites, of "branched thoughts, new grown with pleasant pain" (line 52): "A rosy sanctuary will I dress / With the wreath'd trellis of a working brain" on which the ever-changing unnamable buds of imagination will grow (lines 59–60). The last lines are:

> And there shall be for thee all soft delight
> That shadowy thought can win,
> A bright torch, and a casement ope at night
> To let the warm Love in! (lines 64–67)

A bright light will be cast upon the mind, upon the "wreath'd trellis of a working brain." Psyche, the knowledge of mind, need no longer tremble lest it frighten Cupid (Love) away. The light of knowledge will be there for Love to see; it will be a beacon that invites Love. Inscribed in the body in all its neuroplastic detail—"branched thoughts, new grown with pleasant pain"—the mind will be seen in its full material embodiment, and thought, as it is drawn out of obscurity, will give Psyche herself unending delight. We might remember that the child of Cupid and Psyche was named Pleasure.

The Olympians have grown "faint," but Psyche's "lucent fans" (line 41) (her butterfly wings) radiate light, and Keats' own vision ("by mine own eyes inspired" [line 42]) will lend them the power to fly again, outsoaring the once mighty gods. Such is the power of light; this is "enlightenment" indeed! The confidence that the most obscure recess in the known universe, the source of thought, may be thrown open to view. Psyche is a new goddess, but, rejecting ignorance, she is also a new Eve.

Still, this new goddess of knowledge is a humble, not a prideful, figure. She must be seen and blessed by love. It is only when permeated by love that she can be her divine self. "O for a Life of Sensations rather than of Thoughts!," Keats exclaims (November 22, 1817, letter); yet it is only from within a life of sensations, from the vantage point of love without reservations, that we can do justice to the divinity of thought, that we can restore thought to its natural, divine state. Here Condillac's "Penser est sentir" ["Thinking is feeling"] (quoted in Massey, 1977, p. 330) takes on a new, and better, meaning. Without love, uninspired, thought is sterile. Could Keats have been more explicit? The point of the whole thing is "To let the warm Love in!"

I have said repeatedly that it is only from within the condition of inspiration that we should think about the relation of the brain to poetry, to music, to art. Everything else demeans the effort and dooms it. Keats, at the beginning of the revolution in brain science, saw this truth more clearly than we do now. It is undoubtedly too much to ask of ourselves, but, at least in principle, in order to reveal the proper relation of art and neuroscience, one should be able to express it as a poem.

~~~~~

Bibliography

WORKS CITED

Abbott, H. P. (2006). Cognitive Literary Studies: The "Second Generation." *Poetics Today*, 27(4): 711–722.

Adams, H. ([1907] 1931). *The Education of Henry Adams*. New York: Random House.

Adams, J. (2007). *Interference Patterns: Literary Study, Scientific Knowledge, and Disciplinary Autonomy*. Lewisburg: Bucknell University Press.

Addison, J. ([1711] 1836). *The Spectator*. Two Volumes Complete in One. Philadelphia: J. J. Woodward.

Adorno, T. W. ([2005] 2007). *Dream Notes*. Cambridge, UK: Polity Press.

Agacinski, S., Derrida, J., Kofman, S., Lacoue-Labarthe, P., Nancy, J.-L., and Pautrat, B. (1975). *Mimesis: Des articulations*. Paris: Aubier-Flammarion.

Alexander, M. P. (1997). Aphasia: Clinical and Anatomic Aspects. In T. E. Feinberg and M. J. Farah (Eds.), *Behavioral Neurology and Neuropsychology* (pp. 133–149). New York: McGraw-Hill.

Anderson, S. W. (1981). Vowel Timing and Linguistic Organization of Articulatory Sequences in Jargonaphasia. In J. W. Brown (Ed.), *Jargonaphasia: Perspectives in Neurolinguistics, Neuropsychology, and Psycholinguistics* (pp. 295–304). New York: Academic Press.

Andrews, D. (1976). *The Major Film Theories: An Introduction*. New York: Oxford University Press.

Angier, N. (2001). Sonata for Humans, Birds and Humpback Whales. *New York Times*, January 9, D5.

Anker, S., and Nelkin, D. (2004). *The Molecular Gaze: Art in the Genetic Age*. Cold Spring Harbor, NY: Cold Spring Harbor Laboratory Press.

Antrobus, J. S., and Bertini, M. (Eds.) (1992). *The Neuropsychology of Sleep and Dreaming*. Hillsdale, NJ: Lawrence Erlbaum.

Apfelbach, H. (1922). *Das Denkgefühl: Eine Untersuchung über den emotionalen Charakter der Denkprozesse*. Vienna: Wilhelm Braumüller.

Aristotle (1941). *Poetics*. In R. McKeon (Ed.), *The Basic Works of Aristotle* (pp. 1455–1487). New York: Random House.

Arkin, A. M. (1981). *Sleep-Talking: Psychology and Psychophysiology*. Hillsdale, NJ: Lawrence Erlbaum.

Arnheim, R. (1977). *Dynamics of Architectural Form*. Berkeley: University of   191
California Press.

Arréat, L. (1897). Review of L. Arnoult, *Traité d'esthétique visuelle transcendentale*. *Revue philosophique de la France et de l'étranger*, 22(7): 441–443.

Asher, J. (2005). Gene Knockout Scores a Fearless Mouse. NIH News,
National Institutes of Health, November 22. http://www.nih.gov/news/
pr/nov2005/nimh-22.htm (accessed February 17, 2007).

Augustine of Hippo. (1909). *The Confessions of St. Augustine*. In *The Confessions of St. Augustine*; *The Imitation of Christ by Thomas à Kempis*
(pp. 5–106). New York: P. F. Collier & Son.

Avanzini, G. (2003). Musicogenic Seizures. In G. Avanzini, C. Faienza,
D. Minciacchi, L. Lopez, and M. Majno (Eds.), *The Neurosciences and
Music* (pp. 95–102). New York: New York Academy of Sciences.

Bach-y-Rita, P., Tyler, M. E., and Kaczmarek, K. A. (2003). Seeing with the
Brain. *International Journal of Human-Computer Interaction*, 15(2):
285–295.

Badiou, A. (2003). *Saint Paul: The Foundation of Universalism*. Stanford,
CA: Stanford University Press.

———. (2007). *The Century*. Cambridge, UK: Polity Press.

Bar, M., and Neta, M. (2007). Visual Elements of Subjective Preference
Modulate Amygdala Activation. *Neuropsychologia*, 45(10): 2191–2200.

Barrett, D. L. (2001). *The Committee of Sleep: How Artists, Scientists, and
Athletes Use Dreams for Creative Problem-Solving—And How You Can
Too*. New York: Crown.

Baudelaire, C. ([1861] 1968). "Correspondances." In *Oeuvres Complètes*
(p. 11). Paris: Pléiade.

Baynes, K., and Gazzaniga, M. S. (2005). Lateralization of Language: Toward
a Biologically Based Model of Language. *Linguistic Review*, 22: 303–326.

Beethoven, L. van (1996). *Briefwechsel: Gesamtausgabe*, Vol. 4. Munich: G.
Henle.

Begley, S. (2007). *Train your Mind, Change your Brain: How a New Science
Reveals our Extraordinary Potential to Transform Ourselves*. New York:
Ballantine.

Belin, P., Van Eeckhout, Ph., Zilbovicius, M., Remy, P., Guillaume, S., Chain,
F., Rancurel, G., and Samson, Y. (1996). Recovery from Non-fluent Aphasia after Melodic Intonation Therapy. *Neurology*, 47: 1504–1511.

Benjamin, W. ([1916] 1979). On Language as Such and on the Language
of Man. In *Reflections: Essays, Aphorisms, Autobiographical Writings*
(pp. 314–332). New York: Harcourt Brace Jovanovich.

Bennett, M. R., and Hacker, P. M. S. (2003). *Philosophical Foundations of
Neuroscience*. Malden, MA: Blackwell.

Bentall, R. (2003). *Madness Explained: Psychosis and Human Nature*. London: Penguin Books.

Benton, A. L. (1980). The Amusias. In M. Critchley and R. A. Henson (Eds.),

Music and the Brain: Studies in the Neurology of Music (pp. 378–397).
London: William Heinemann.

Benton, A. L., and Joynt, R. J. (1960). Early Descriptions of Aphasia.
Archives of Neurology, 3: 205–222.

Benzon, W. L. (2001). *Beethoven's Anvil: Music in Mind and Culture*. New
York: Basic Books.

Bergson, H. ([1888] 1945). *Essai sur les données immédiates de la conscience*.
Geneva: Alber Skira.

Bersani, L. (1981). Representation and its Discontents. In S. Greenblatt (Ed.),
Allegory and Representation (pp. 145–162). Baltimore: Johns Hopkins
University Press.

Bertini, M. (1984). Il linguaggio del sogno attraverso il sonno e la veglia.
In V. Branca, C. Ossola, and S. Resnick (Eds.), *I linguaggi del sogno*
(pp. 251–266). Firenze: Sansoni.

Bickle, J. W. (2003). Empirical Evidence for a Narrative Concept of Self. In
G. D. Fireman, T. E. McVay, Jr., and O. J. Flanagan (Eds.), *Narrative and
Consciousness: Literature, Psychology, and the Brain* (pp. 195–208). New
York: Oxford University Press.

Bishop, E. ([1946] 1977). "The Man-Moth." In *Elizabeth Bishop: The Complete Poems* (pp. 15–16). New York: Farrar, Straus and Giroux.

———. "One Art." In *Elizabeth Bishop: The Complete Poems* (p. 178). New
York: Farrar, Straus and Giroux.

Blakeslee, S. (1995). Music in the Brain: Is Ability to Perceive, Enjoy a Tune
an Inborn Human Trait? *Houston Chronicle*, May 22, A8.

Blechner, M. J. (2001). *The Dream Frontier*. Hillsdale, NJ: Analytic Press.

Blum, D. (1996). The Healing Power of Music. *BBC Music Magazine*, April,
28–30.

Blumenberg, H. (1974). *Säkularisierung und Selbstbehauptung*. Frankfurt:
Suhrkamp.

Bogousslavsky, J., and Boller, F. (Eds.) (2005). *Neurological Disorders in
Famous Artists*. Basel: Karger.

Bogousslavsky, J., and Hennerici, M. G. (Eds.) (2007). *Neurological Disorders in Famous Artists, Part 2*. Basel: Karger.

Boon, M. (2002). *The Road of Excess: A History of Writers on Drugs*.
Cambridge, MA: Harvard University Press.

Borges, J. L. (1962). "The Circular Ruins." In *Ficciones* (pp. 57–63). New
York: Grove Press.

Boucher, V., Garcia, L. J., Fleurant, J., and Paradis, J. (2001). Variable Effi-
cacy of Rhythm and Tone in Melody-Based Interventions: Implications
for the Assumption of a Right-Hemisphere Facilitation in Non-Fluent
Aphasia. *Aphasiology*, 15: 131–149.

Breton, A. (1962). *Manifestes du surréalisme*. Paris: Jean-Jacques Pauvert.

Brooks, M. (2007). The Universe Machine. *New Scientist*, March 30: 30–34.

Brown, J. W. (1979). *Aphasia, Apraxia and Agnosia: Clinical and Theoretical
Aspects*. Springfield, IL: Charles C. Thomas.

———. (Ed.) (1981). *Jargonaphasia: Perspectives in Neurolinguistics, Neuro-*
psychology, and Psycholinguistics. New York: Academic Press.

———. (1999). On Aesthetic Perception. *Journal of Consciousness Studies,* 6: 143–161.

Brown, S. (2000). The "Musilanguage" Model of Music Evolution. In N. L. Wallin, B. Merker, and S. Brown (Eds.), *The Origins of Music* (pp. 271–314). Cambridge, MA: MIT Press.

———. (2006). The Perpetual Music Track. *Journal of Consciousness Studies,* 13(6): 25–44.

———. (2008). The Neuroscience of Dance. *Scientific American,* July, 78–83.

Brust, J. C. M. (2001). Music and the Neurologist: A Historical Perspective. In R. J. Zatorre and I. Peretz (Eds.), *The Biological Foundations of Music* (pp. 143–152). New York: New York Academy of Sciences.

Byrne, R. M. J. (2005). *The Rational Imagination: How People Create Alternatives to Reality.* Cambridge, MA: MIT Press.

Cadalbert, A., Landis, T., Regard, M., and Graves, R. E. (1994). Singing With and Without Words: Hemispheric Asymmetries in Motor Control. *Journal of Clinical and Experimental Neuropsychology,* 16: 664–670.

Calame, G. (1878). Review of T. Fechner, *Vorschule der Aesthetik. Revue philosophique de la France et de l'étranger,* 3(8): 172–189.

Carey, B. (2005). After a Stroke, Artistic Transformation. *New York Times,* May 24, F5.

Carey, J. (2005). *What Good are the Arts?* London: Faber and Faber.

Carrol, L. (1960). *The Annotated Alice.* New York: Meridian.

Cela-Conde, C. J., Marty, G., Maestu, F., Ortiz, T., Munar, E., Fernandez, A., Roca, M., Rossello, J., and Quesney, F. (2004). Activation of the Prefrontal Cortex in the Human Visual Aesthetic Perception. *Publications of the National Academy of Sciences,* 101: 6321–6325.

Changeux, J.-P. (2004). *The Physiology of Truth: Neuroscience and Human Knowledge.* Cambridge, MA: Harvard University Press.

Cohen, R. (2001). *Inspired Sleep.* New York: Scribner.

Coleridge, S. T. ([1816] 1957). "Christabel." In *The Complete Poetical Works of Samuel Taylor Coleridge,* Vol. 1 (pp. 213–236). Oxford: Clarendon Press.

———. ([1816] 1957). "Kubla Khan." In *The Complete Poetical Works of Samuel Taylor Coleridge,* Vol. 1 (pp. 297–298). Oxford: Clarendon Press.

———. (1957). *The Notebooks of Samuel Taylor Coleridge,* Vol. 1. Princeton: Princeton University Press.

Configurations (2002). 10(3).

Corballis, M. C. (1999). The Gestural Origin of Language. *American Scientist,* 87: 138–145.

Crane, M. T. (2001). *Shakespeare's Brain: Reading with Cognitive Theory.* Princeton: Princeton University Press.

Crary, J. (1990). *Techniques of the Observer: On Vision and Modernity in the Nineteenth Century.* Cambridge, MA: MIT Press.

Cuillé, T. B. (2006). *Narrative Interludes: Musical Tableaux in Eighteenth-Century French Texts*. Toronto: University of Toronto Press.

Currie, J. R. (2002). Better the Puppet? *Current Musicology*, 74(Fall): 5–45.

Danto, A. (2003). *The Abuse of Beauty: Aesthetics and the Concept of Art*. Chicago: Open Court Publishing Company.

Darwin, C. ([1871] 1902). *Descent of Man and Selection in Relation to Sex*. 2 Vols. New York: P. F. Collier and Son.

Dave, A. S., Yu, A. C., and Margoliash, D. (1998). Behavioral State Modulation of Auditory Activity in a Vocal Motor System. *Science*, 282: 2250–2254.

Davidson, R. J. (1995). Cerebral Asymmetry, Emotion, and Affective Style. In R. J. Davidson and K. Hugdahl (Eds.), *Brain Asymmetry* (pp. 361–387). Cambridge, MA: MIT Press.

Davies, A. (1984). Private Enigma in the Opened Text. In B. Andrews and C. Bernstein (Eds.), *The L=A=N=G=U=A=G=E Book* (pp. 7–10). Carbondale: Southern Illinois University Press.

Davis, Philip (2007). The Shakespeared Brain. *Moreintelligentlife.com*. http://moreintelligentlife.com/node/298/ (accessed December 9, 2007).

Deacon, T. W. (1997). *The Symbolic Species: The Co-Evolution of Language and the Brain*. New York: W. W. Norton.

Decety, J., Michalska, K. J., and Akitsuki, Y. (2008). Who Caused the Pain? An fMRI Investigation of Empathy and Intentionality in Children. *Neuropsychologia* 46(11): 2607–2614.

Dennett, D. (2005). *Sweet Dreams: Philosophical Obstacles to a Science of Consciousness*. Cambridge, MA: MIT Press.

Derrida, J. (1971). La mythologie blanche. *Poétique* 5: 1–52.

Dieguez, S., Assal, G., and Bogousslavsky, J. (2007). Visconti and Fellini: From Left Social Neorealism to Right-Hemisphere Stroke. In J. Bogousslavsky and M. G. Hennerici (Eds.), *Neurological Disorders in Famous Artists* (pp. 44–74). Basel: Karger.

Dilthey, W. ([1883] 1961). *Einleitung in die Geisteswissenschaften*. Gesammelte Schriften, Vol. 1. Göttingen: Vandenhoeck und Ruprecht.

DNA variety items for sale [Advertisement] (2006). *Discover*, April.

Doidge, N. (2007). *The Brain that Changes Itself: Stories of Personal Triumph from the Frontiers of Brain Science*. New York: Viking Adult.

Draaisma, D. (2000). *Metaphors of Memory: A History of Ideas about the Mind*. Cambridge, UK: Cambridge University Press.

———. (2004). *Why Life Speeds up as you get Older: How Memory Shapes our Past*. Cambridge, UK: Cambridge University Press.

Dugas, L. (1905). Sur les abstraits émotionnels. *Revue philosophique de la France et de l'étranger*, 30(11): 472–485.

Dunsby, J. (1992). *Schoenberg: Pierrot Lunaire*. New York: Cambridge University Press.

Du Plessis, R. B. (2008). "Some Codas." Holiday Broadside. The Poetry Collection, State University of New York at Buffalo.

Eberhard, J. (2007). *Architecture and the Brain: A New Knowledge Base from Neuroscience.* Atlanta: Greenway Communications.

Ede, S. (2005). *Art and Science.* London: I. B. Tauris.

Edelman, G. M. (2004). *Wider than the Sky: The Phenomenal Gift of Consciousness.* New Haven: Yale University Press.

———. (2006). *Second Nature: Brain Science and Human Knowledge.* New Haven: Yale University Press.

Edwards, E. (2008). *Artscience: Creativity in the Post-Google Generation.* Cambridge, MA: Harvard University Press.

Efron, R. (1990). *The Decline and Fall of Hemispheric Specialization.* Hillsdale, NJ: Lawrence Erlbaum.

Elfenbein, A. (2006). Cognitive Science and the History of Reading. *PMLA,* 121(2): 484–502.

Eliot, G. (1885). *Poems together with Brother Jacob and the Lifted Veil.* New York: Harper and Brothers.

Elkins, J. (2008). *Six Stories from the End of Representation: Images in Painting, Photography, Astronomy, Microscopy, Particle Physics, and Quantum Mechanics, 1980–2000.* Stanford: Stanford University Press.

Evenson, B. (2003). Why We Can't Get that Song Out of Our Heads. *National Post,* July 7, A1 and A14.

Faber, R., Abrams, R., Taylor, M. A., Kasprison, A., Morris, C., and Weisz, R. (1983). Comparison of Schizophrenic Patients with Formal Thought Disorder and Neurologically Impaired Patients with Aphasia. *American Journal of Psychiatry,* 140(10): 1348–1351.

Faienza, C., and Cossu, G. (2003). Introductory Remarks on Musical Beginnings: Ten Years Later. In J. Avanzini, C. Faienza, D. Minciacchi, L. Lopez, and M. Majno (Eds.), *The Neurosciences and Music* (pp. 392–396). New York: New York Academy of Sciences.

Fermor, P. L. (1977). *A Time of Gifts.* London: Penguin Books.

———. (1987). *Between the Woods and the Water.* London: Penguin Books.

Fernald, A. (1989). Intonation and Communicative Intent in Mothers' Speech to Infants: Is the Melody the Message? *Child Development,* 60(6): 1497–1510.

Fiedler, K. (1971). *Schriften zur Kunst.* 2 Vols. Munich: Wilhelm Fink.

Flaherty, A. (2004). *The Midnight Disease: The Drive to Write, Writer's Block, and the Creative Brain.* Boston: Houghton Mifflin.

Foulkes, D. (1985). *Dreaming: A Cognitive-Psychological Analysis.* Hillsdale, NJ: Lawrence Erlbaum.

———. (1999). *Children's Dreaming and the Development of Consciousness.* Cambridge, MA: Harvard University Press.

Fox, D. (2007). Consciousness in a . . . Cockroach? *Discover,* January, 67–70.

Franzen, J. (2001). *The Corrections.* New York: Picador.

Freeman, W. (2000). A Neurobiological Role of Music in Social Bonding. In

N. L. Wallin, B. Merker, and S. Brown (Eds.), *Origins of Music* (pp. 411–424). Cambridge, MA: MIT Press.

Freud, S. (1900). *Die Traumdeutung*. Leipzig: Franz Deuticke.

Fried, I., Wilson, C., MacDonald, K. A., and Behnke, E. J. (1998). Electric Current Stimulates Laughter. *Nature*, 391: 650.

Gamwell, L. (2002). *Exploring the Invisible: Art, Science, and the Spiritual*. Princeton: Princeton University Press.

Garfield, K. (2006). Are We all Synesthetes? *Discover*, December, 19.

Gauss, C. E. (1949). *The Aesthetic Theories of French Artists 1855 to the Present*. Baltimore: Johns Hopkins University Press.

Gazzaniga, M. S. (2005). *The Ethical Brain*. New York: Dana Press.

Geschwind, N. (1965). Disconnexion Syndromes in Animals and Man, Part 2. *Brain*, 88: 585–644.

Ghyka, M. (1952). *A Practical Handbook of Geometrical Composition and Design*. London: Alec Tiranti.

Glass, P. (2006). Lecture for the Inauguration of the Center for New Music, University of Buffalo, September 18.

Glazier, L. (2002). *Digital Poetics: The Making of E-Poetries*. Tuscaloosa: University of Alabama Press.

Glucksberg, S. (2001). *Understanding Figurative Language: From Metaphors to Idioms*. New York: Oxford University Press.

Goel, V., and Dolan, R. J. (2001). The Functional Anatomy of Humor. *Nature Neuroscience*, 4(3): 237–238.

Goldman, A. J. (2006). *Simulating Minds: The Philosophy, Psychology, and Neuroscience of Mindreading*. New York: Oxford University Press.

Goodale, M. A., and Milner, A. D. (2004). *Sight Unseen: An Exploration of Conscious and Unconscious Vision*. New York: Oxford University Press.

Gopnik, A. (2007). Cells that Read Minds? *Slate*, April 26. http://www .slate.com/id/2165123/ (accessed November 22, 2008).

Grace, N. (2002). Making Dreams into Music: Contemporary Songwriters Carry On an Age-Old Dreaming Tradition. Paper read at a conference of the Association for the Study of Dreams at Tufts University, Medford, MA, June 18.

Graham, S. (2005). Brain Region Linked to Metaphor Comprehension. *SciAm.com*, May 26. http://www.sciam.com/article.cfm?id=brain-region -linked-to-me (accessed December 17, 2008).

Gravitz, L. (2006). Rain Man's Brain Explained. *Discover*, April, 17.

Gray, P. M., Krause, B., Atema, J., Payne, R., Krumhansi, C., and Baptista, L. (2001). The Music of Nature and the Nature of Music. *Science*, 291: 52–54.

Greenberg, M. (2008). Remember This. *New York Review of Books*, December 4, 10–14.

Greenberg, N. (2004). The Beast at Play: The Neuroethology of Creativity. In R. Clements and L. Fiorentino (Eds.), *The Child's Right to Play: A Global Approach* (pp. 309–327). Westport, CT: Praeger.

Grieco, P. (2007). Postscript. *The Poetry Revolt*, 1(1): 3.

Gross, D. M. (2006). *The Secret History of Emotion: From Aristotle's "Rhetoric" to Modern Brain Science*. Chicago: University of Chicago Press.

Gustafson, R. (1986). *Leo Tolstoy, Resident and Stranger: A Study in Fiction and Theology*. Princeton: Princeton University Press.

Hacking, I. (1995). *Rewriting the Soul: Multiple Personality and the Sciences of Memory*. Princeton: Princeton University Press.

Haddon, M. (2003). *The Curious Incident of the Dog in the Night-Time*. New York: Doubleday.

Halmi, N. (2007). *The Genealogy of the Romantic Symbol*. New York: Oxford University Press.

Hamilton, J. T. (2008). *Music, Madness, and the Unworking of Language*. New York: Columbia University Press.

Hardy, L. (2007). Quantum Gravity Computers: On the Theory of Computation with Indefinite Causal Structure. arvix.org, Cornell University Library. http://arxiv.org/abs/quant-ph/0701019 (accessed June 29, 2007).

Harmon-Jones, E. (2003). Clarifying the Emotive Functions of Asymmetrical Frontal Cortical Activity. *Psychophysiology*, 40(6): 838–848.

Harrington, A. (1987). *Medicine, Mind, and the Double Brain: A Study in Nineteenth-Century Thought*. Princeton: Princeton University Press.

Harrison, C. (2001). *Conceptual Art and Painting: Further Essays on Art and Language*. Cambridge, MA: MIT Press.

Haseltine, E. (2003). Round and Round. *Discover*, May, 88.

Hauser, M. D. (2006). *Moral Minds: How Nature Designed our Universal Sense of Right and Wrong*. New York: Ecco.

Hawkins, J., and Blakeslee, S. (2004). *On Intelligence*. New York: Henry Holt.

Hébert, S., and Peretz, I. (2001). Are Text and Tune of Familiar Songs Separable by Brain Damage? *Brain and Cognition* 46: 169–175.

Hébert, S., Racette, A., Gagnon, L., and Peretz, I. (2003). Revisiting the Dissociation between Singing and Speaking in Expressive Aphasia. *Brain*, 126: 1838–1850.

Heidegger, M. ([1927] 1967). *Sein und Zeit*. Tübingen: Max Niemeyer.

Heider, F., and Simmel, M. (1944). An Experimental Study of Apparent Behavior. *American Journal of Psychology*, 57: 243–259.

Heilman, K. M. (2005). *Creativity and the Brain*. New York: Psychology Press.

Herling, G. (1967). *The Island: Three Tales*. Cleveland: World Publishing.

Herzog, G. (1936). A Comparison of Pueblo and Pima Musical Styles. *Journal of American Folklore*, 49: 283–417.

Heynick, F. (1993). *Language and its Disturbances in Dreams: The Pioneering Work of Freud and Kraepelin Updated*. New York: John Wiley and Sons.

Hilferty, R. (2006). Mozart's Librettist Gets the Stage. *Forward*, August 25, 13.

Hirsch, G., and Hirsch, L. (2006). Trollope's *The Last Chronicle of Barset*: Memory, Depression, and Cognitive Science. *Mosaic* 39(1): 165–179.

Hobson, J. A. (1988). *The Dreaming Brain*. New York: Basic Books.

Hobson, J. A., Stickgold, R., and Pace-Schott, E. (1998). The Neuro-psychology of REM Sleep Dreaming. *NeuroReport*, 9(3): R1–R14.

Hoffmann, E. T. A. (n. d.). *Musikalische Novellen und Aufsätze*. Leipzig: Insel.

Hogan, P. C. (2003). *Cognitive Science, Literature, and the Arts: A Guide for Humanists*. New York: Routledge.

Hopkins, G. M. (1959). *The Journals and Papers of Gerard Manley Hopkins*. New York: Oxford University Press.

———. (1967). "The Habit of Perfection." In *The Poems of Gerard Manley Hopkins* (pp. 31–32). New York: Oxford University Press.

Howe, S. (2002). From "Pythagorean Silence." In R. Silliman (Ed.), *In the American Tree* (p. 347). Orono, ME: National Poetry Foundation.

Huhn, T. (2004). *Imitation and Society: The Persistence of Mimesis in the Aesthetics of Burke, Hogarth, and Kant*. University Park: Pennsylvania State University Press.

Humphrey, N. (2006). *Seeing Red: A Study in Consciousness*. Cambridge, MA: Belknap Press.

Hunt, H. T. (1989). *The Multiplicity of Dreams: Memory, Imagination, and Consciousness*. New Haven: Yale University Press.

———. (1995). *On the Nature of Consciousness*. New Haven: Yale University Press.

———. (2000). New Multiplicities of Dreaming and REMing. *Behavioral and Brain Sciences*, 23: 953–955.

Hurlburt, R. T., and Schwitzgebel, E. (2007). *Describing Inner Experience? Proponent meets Skeptic*. Cambridge, MA: MIT Press.

Huron, D. (2003). Is Music an Evolutionary Adaptation? In I. Peretz and R. Zatorre (Eds.), *The Cognitive Neuroscience of Music* (pp. 57–75). New York: Oxford University Press.

———. (2006). *Sweet Anticipation: Music and the Psychology of Expectation*. Cambridge, MA: MIT Press.

Iacoboni, M. (2008). *Mirroring People: The New Science of How We Connect with Others*. New York: Farrar, Straus and Giroux.

Jackson, J. H. ([1874] 1958). *Selected Writings*. 2 Vols. New York: Basic Books.

Jackson, N. (2008). *Science and Sensation in Romantic Poetry*. Cambridge, UK: Cambridge University Press.

Jaynes, J. (2000). *The Origin of Consciousness in the Breakdown of the Bicameral Mind*. Boston: Houghton Mifflin.

Jeffries, K. J., Fritz, J. B., and Braun, A. R. (2003). Words in Melody: An H2 15O PET Study of Brain Activation during Singing and Speaking. *Neuroreport*, 14(5): 749–754.

Johnson, E. (2002). The Laughter Circuit. *Discover*, May, 24–25.

Johnson, S. (2003). Emotions and the Brain: Love. *Discover*, May, 71–76.

Jones, C. (2006). *Sensorium: Embodied Experience, Technology, and Contemporary Art*. Cambridge, MA: MIT Press.

Jourdain, R. (1997). *Music, the Brain, and Ecstasy: How Music Captures our Imagination*. New York: William Morrow.

Jouvet, M. (1999). *The Paradox of Sleep: The Story of Dreaming*. Cambridge, MA: MIT Press.

Kafka, F. ([1915] 1970). "Die Verwandlung." In *Sämtliche Erzählungen* (pp. 56–99). Frankfurt: Fischer Bücherei.

Kandel, E. R. (2006). *In Search of Memory: The Emergence of a New Science of Mind*. New York: W. W. Norton.

Karmel, P. (2003). *Picasso and the Invention of Cubism*. New Haven: Yale University Press.

Kawabata, H., and Zeki, S. (2004). Neural Correlates of Beauty. *Neurophysiology*, 91: 1699–1705.

Kay, K. (1991). *The Little Giant Book of Optical Illusions*. New York: Sterling Publishing.

Kayumov, L., Soare, K., Serbine, O., Kotlyar, B., Simkhovich, Z., Goldstein, Y., Gavrilov, D., Levin, Y., and Shapiro, C. M. (2002). Brain Music Therapy for Treatment of Insomnia and Anxiety. *Sleep* 25: A241.

Keats, J. ([1817] 1959). Letter to Benjamin Bailey. In *John Keats: Selected Poems and Letters* (pp. 257–259). Boston: Houghton Mifflin.

———. ([1820] 1959). "Hyperion: A Fragment." In *John Keats: Selected Poems and Letters* (pp. 161–180). Boston: Houghton Mifflin.

———. ([1820] 1959). "Ode on a Grecian Urn." In *John Keats: Selected Poems and Letters* (pp. 207–208). Boston: Houghton Mifflin.

———. ([1820] 1959). "Ode to Psyche." In *John Keats: Selected Poems and Letters* (pp. 203–205). Boston: Houghton Mifflin.

Keijzer, F. (2001). *Representation and Behavior*. Cambridge, MA: MIT Press.

Kenneally, C. (2006). The Deepest Cut. *The New Yorker*, July 3, 36–42.

Kertesz, A. (1981). The Anatomy of Jargon. In J. W. Brown (Ed.), *Jargonaphasia: Perspectives in Neurolinguistics, Neuropsychology, and Psycholinguistics* (pp. 63–112). New York: Academic Press.

Kierkegaard, S. ([1841] 1968). *The Concept of Irony with Constant Reference to Socrates*. Bloomington: Indiana University Press.

Kiš, D. (1990). *Hourglass*. New York: Farrar, Straus and Giroux.

Klawans, H. (2000). *Defending the Cave-Woman and other Tales of Evolutionary Neurology*. New York: W. W. Norton.

Kluger, A. N., Siegfried, Z., and Ebstein, R. P. (2002). A Meta-Analysis of the Association between DRD4 Polymorphism and Novelty Seeking. *Molecular Psychiatry*, 7: 712–717.

Knezovich, M. (2006). I Think, therefore I Pong. *University of Chicago Magazine*, June, 20–21.

Knobloch, F., Postolka, M., and Srnec, J. (1964). Musical Experience as Interpersonal Process. *Psychiatry*, 27: 259–265.

Kofman, S. (1971). L'oubli de la métaphore. *Critique*, 27(291–292): 783–804.

Kolb, B., and Whishaw, I. Q. (2001). *Introduction to Brain and Behavior*. New York: Worth.

Korsmeyer, C. (1998). *Aesthetics: The Big Questions*. Malden, MA: Blackwell.

Kosfeld, M., Heinrichs, M., Zak, P. J., Fischbacher, U., and Fehr, E. (2005). Oxytocin Increases Trust in Humans. *Nature*, 453: 673–676.

Kraepelin, E. (1906). *Über Sprachstörungen im Traume*. Leipzig: Wilhelm Engelmann.

Kruglinski, S. (2008). Musical Scales Mimic Sound of Language. *Discover*, January, 59.

La Berge, S. (1992). Physiological Studies of Lucid Dreaming. In Antrobus and Bertini (Eds.), *The Neuropsychology of Sleep and Dreaming* (pp. 289–303). Hillsdale, NJ: Lawrence Erlbaum.

Labrousse, R. (2004). What Remains Belongs to God. In A. Dumas (Ed.), *Matisse, His Art and His Textiles: The Fabric of Dreams* (pp. 46–61). London: Royal Academy of Arts.

Lacoue-Labarthe, P. (1987). *La Fiction du politique: Heidegger, l'art et la politique*. Paris: Christian Bourgois.

Lacoue-Labarthe, P., and Nancy, J.-L. (1978). *L'absolu littéraire: théorie de la littérature du romantisme allemand*. Paris: Seuil.

Lechevalier, B. (1997). La Perception des sons musicaux: apports de la camera à positons. *Bulletin de l'académie nationale de médecine*, 181: 1191–1199.

Lehrer, J. (2007). *Proust was a Neuroscientist*. New York: Houghton-Mifflin.

Leibniz, G. W. (1900). *Oeuvres philosophiques*. 2 Vols. Paris: Félix Alcan.

Lem, S. (1994). *Peace on Earth*. New York: Harcourt Brace.

Lerdahl, F. (2001). The Sounds of Poetry Viewed as Music. In R. J. Zatorre and I. Peretz (Eds.), *The Biological Foundations of Music* (pp. 337–354). New York: New York Academy of Sciences.

Lethem, J. (1999). *Motherless Brooklyn*. New York: Doubleday.

Levinas, E. ([1948] 1985). *Le Temps et l'autre*. Paris: Presses Universitaires de France.

Levitin, D. J. (2006). *This is your Brain on Music: The Science of a Human Obsession*. New York: Dutton.

———. (2008). *The World in Six Songs: How the Musical Brain Created Human Nature*. New York: Dutton.

Libet, B. (2004). *Mind Time: The Temporal Factor in Consciousness*. Cambridge, MA: Harvard University Press.

Limb, C. J., and Braun, A. R. Neural Substrates of Spontaneous Musical Performance: An fMRI Study of Jazz Improvisation. PLoS ONE. http://www.plosone.org/article/info:doi%2F10.1371%2Fjournal.pone.0001679 (accessed December 17, 2008).

Limosin, F., Loze, J.-Y., Rouillon, F., Adès, J., and Gorwood, P. (2003). Association between Dopamine Receptor D1 Gene DdeI Polymorphism and Sensation Seeking in Alcohol-Dependent Men. *Alcoholism: Clinical and Experimental Research*, 27(8): 1226–1228.

Lipps, T. (1891). Aesthetische Faktoren der Raumanschauung. In T. W. Engelmann, E. Javal, A. Koenig, J. von Kries, T. Lipps, L. Matthiessen, W. Preyer, and W. Uhthoff (Eds.), *Beiträge zur Psychologie und Physiolo-*

gie der Sinnesorgane: Hermann von Helmholtz als Festgruss zu seinem siebzigsten Geburtstag (pp. 217–307). Hamburg: Leopold Voss.

———. ([1904] 1909). Ueber einfachste Formen der Raumkunst (with a supplement of five plates). *Abhandlungen der philosophisch-philologischen Klasse der königlich bayerischen Akademie der Wissenschaften*, 23: 399–480.

———. (1907). Das Wissen von fremden Ichen. *Psychologische Untersuchungen*, 1: 694–722.

Livingston, I. (2006). *Between Science and Culture: An Introduction to Autopoetics*. Urbana: University of Illinois Press.

Livingstone, M. (2002). *Vision and Art: The Biology of Seeing*. New York: Harry N. Abrams.

Llinás, R. R., and Paré, D. (1991). Of Dreaming and Wakefulness. *Neuroscience*, 44(3): 521–535.

Lloyd, D. (2004). *Radiant Cool: A Novel Theory of Consciousness*. Cambridge, MA: MIT Press.

Lythgoe, M. F., Pollak, T. A., Kalmas, M., de Haan, M., and Chong, W. (2005). Obsessive, Prolific Artistic Output Following Subarachnoid Hemorrhage. *Neurology*, 64: 397–398.

Macrae, F. (2008). The 'thinking cap' that could unlock your inner genius and boost creativity. *Mail Online*, September 30. http://www.dailymail .co.uk/sciencetech/article-1064431/The-thinking-cap-unlock-inner -genius-boost-creativity.htm (accessed November 16, 2008).

Mallgrave, H. F., and Ikonomou, E. (1994). *Empathy, Form, and Space: Problems in German Aesthetics 1873–1893*. Santa Monica: Getty Center for the History of Art and the Humanities.

Margalit, A. (2005). The Genius of Spinoza. *New York Review of Books*, October 20, 48–50.

Markoff, J. (2005). Researchers Look to Create a Synthesis of Art and Science for the 21st Century. *New York Times*, November 5, A17.

Martin, R. C., and Freedman, M. L. (2001). Relations between Language and Memory Deficits. In R. S. Berndt (Ed.), *Language and Aphasia* (pp. 239–256). Handbook of Neuropsychology, 2nd ed., Vol. 3. Amsterdam: Elsevier.

Martindale, C. (2007). A Neural-Network Theory of Beauty. In C. Martindale, P. Locher, and V. Petrov (Eds.) *Evolutionary and Neurocognitive Approaches to Aesthetics, Creativity and the Arts* (pp. 181–194). Amityville, NY: Baywood.

Massey, I. (1954). The Contribution of Neurology to the Scepticism of Alfred de Vigny. *Journal of the History of Medicine and Allied Sciences*, 9: 329–348.

———. (1976). *The Gaping Pig: Literature and Metamorphosis*. Berkeley: University of California Press.

———. (1977). Two Types of Visual Metaphor. *Criticism*, 19: 285–295.

———. (1987a). *Find You the Virtue: Ethics, Image, and Desire in Literature*. Fairfax, VA: George Mason University Press.

———. (Ed.). (1987b). *Bodleian MS. Shelley adds. d. 7: A Facsimile Edition with Full Transcription and Textual Notes.* New York: Garland.

———. (1990). Freud before Freud: K. A. Scherner (1825–1889). *Centennial Review,* 34(4): 567–576.

———. (1991). Notes on the Imagination. In P. Baker, S. W. Goodwin, and G. Handwerk (Eds.), *The Scope of Words* (pp. 29–38). New York: Peter Lang.

———. (1999). Notes on the Imagination, Part III. *Paunch,* 69–70: 12–21.

———. (2006a). The Musical Dream Revisited: Music and Language in Dreams. *Psychology of Aesthetics, Creativity, and the Arts,* S1: 42–50.

———. (2006b). The Wounded Body in Canadian and American Yiddish Poetry. In A. Preis-Smith, E. Luczak, and M. Paryz (Eds.), *Mosaics of Words* (pp. 33–43). Warsaw: University of Warsaw Press.

Mattheson, J. (1739). *Der vollkommene Kapellmeister.* Hamburg: Christian Herold.

Matthews, P. M., and McQuain, J. (2003). *The Bard on the Brain: Understanding the Mind through the Art of Shakespeare and the Science of Brain Imaging.* New York: Dana Press.

Mazurkewich, K. (2006). 33,300 Songs and Still Writing. *Toronto Globe and Mail,* July 15, R1 and R7.

McElroy, S. (2006). What's So Funny about Tourette's? *New York Times,* January 15, Section 2, 5.

McEwan, I. (2006). *Saturday.* London: Vintage.

McKeon, R. (Ed.) (1941). *The Basic Works of Aristotle.* New York: Random House.

McNamara, P., Andresen, J., Arrowood, J., and Messer, G. (2002). Counterfactual Cognitive Operations in Dreams. *Dreaming,* 12: 121–133.

Melser, D. (2004). *The Act of Thinking.* Cambridge, MA: MIT Press.

Merleau-Ponty, M. (1969). *The Essential Writings of Merleau-Ponty* (A. L. Fisher, Ed.). New York: Harcourt, Brace and World.

Meyer, S. (2001). *Irresistible Dictation: Gertrude Stein and the Correlations of Writing and Science.* Stanford: Stanford University Press.

Miller, G. A. (1962). *Psychology: The Science of Mental Life.* New York: Harper and Row.

Miller, M. (1987). Musician's Reputation Began to Spread Only when He was 62. *Toronto Globe and Mail,* July 17, D11.

Minciacchi, D. (2003). Translation from Neurobiological Data to Music Parameters. In J. Avanzini, C. Faienza, D. Minciacchi, L. Lopez, and M. Majno (Eds.), *The Neurosciences and Music* (pp. 282–301). New York: New York Academy of Sciences.

Minkel, J. R. (2006). Brain Damage Can Make You Brilliant. *Discover,* October, 18.

Mitchell, J. P., Mahzarin, R. B., and Macrae, C. N. (2005). The Link between Social Cognition and Self-Referential Thought in the Medial Prefrontal Cortex. *Journal of Cognitive Neuroscience,* 17(8): 1306–1315.

Mithen, S. (2005). *The Singing Neanderthals: The Origins of Music, Language, Mind and Body*. London: Weidenfeld and Nicolson.

Moberg, K. U. (2003). *The Oxytocin Factor*. Cambridge, MA: Da Capo Press.

Modell, A. H. (2003). *Imagination and the Meaningful Brain*. Cambridge, MA: MIT Press.

Molino, J. (2000). Toward an Evolutionary Theory of Music and Language. In N. L. Wallin, B. Merker, and S. Brown (Eds.), *The Origins of Music* (pp. 165–176). Cambridge, MA: MIT Press.

Molnar-Szakacs, I., and Overy, K. (2006). Music and Mirror Neurons: From Motion to "e"motion. *Social, Cognitive, and Affective Neuroscience*, 1(3): 235–241.

Morris, C. (1971). Esthetics and the Theory of Signs. In T. A. Sebeok (Ed.), *Writings on the General Theory of Signs* (pp. 415–433). The Hague: Mouton.

Morrison, T. (2004). *Beloved*. New York: Vintage.

Moscovitch, M. (1973). Language and the Cerebral Hemispheres: Reaction-Time Studies and their Implications for Models of Cerebral Dominance. In P. Pliner, L. Kramer, and T. Alloway (Eds.), *Communication and Affect: Language and Thought* (pp. 89–126). New York: Academic Press.

Music on the Brain (2001). *Wellcome News*, 27, Q2, 24–25.

Nabokov, V. (1966). *Speak, Memory*. New York: G. P. Putnam's Sons.

Nagourney, E. (2001). A Song in your Head Can Turn Deadly. *New York Times*, April 24, F8.

Nalbantian, S. (2003). *Memory in Literature: From Rousseau to Neuroscience*. Houndmills, UK: Palgrave Macmillan.

Nancy, J.-L. (2003). *Au Fond des images*. Paris: Galilée.

Nettl, B. (1986). History and Change in Blackfoot Indian Musical Culture and Thought. *The World of Music*, 28: 70–84.

Noë, A. (2004). *Action in Perception*. Cambridge, MA: MIT Press.

Oates, J. C. (2007). Lest we Forget. *New York Review of Books*, July 19, 47–50.

O'Gorman, M. (2006). *E-Crit: Digital Media, Critical Theory, and the Humanities*. Toronto: University of Toronto Press.

Oke, A. F., Carver, L. A., and Adams, R. N. (1993). Dopamine-Initiated Disturbances of Thalamic Information Processing in Schizophrenia? In R. L. Cromwell and C. R. Snyder (Eds.), *Schizophrenia: Origins, Processes, Treatment, and Outcome* (pp. 31–47). New York: Oxford University Press.

Origins of the brain: Complex synapses drove brain evolution. (2008). Sanger Institute, June 8. http://www.sanger.ac.uk/Info/Press/2008/080608.shtml (accessed August 6, 2008).

Orr, H. A. (2003). What's Not in your Genes. *New York Review of Books*, August 14, 1438–40.

Osborne, L. (2003). Savant for a Day. *New York Times Magazine*, June 22, 38–41.

Ostrow, J. M. (1990). *Social Sensitivity: A Study of Habit and Experience*. Albany: State University of New York Press.

Oswald, I. (1962). *Sleeping and Waking: Physiology and Psychology*. Amsterdam: Elsevier.

Otis, L. (1999). *Membranes: Metaphors of Invasion in Nineteenth-Century Literature, Science, and Politics*. Baltimore: Johns Hopkins University Press.

———. (2001). *Networking: Communicating with Bodies and Machines in the Nineteenth Century*. Ann Arbor: University of Michigan Press.

Panksepp, J. (1998). *Affective Neuroscience: The Foundations of Human and Animal Emotions*. New York: Oxford University Press.

Patel, A. D. (2003). Language, Music, Syntax and the Brain. *Nature Neuroscience*, 6: 674–681.

———. (2008). *Music, Language, and the Brain*. New York: Oxford University Press.

Paustovsky, K. (1986). *In the Heart of Russia and Other Stories*. Moscow: Raduga.

Peacock, T. L. ([1820] 1965). The Four Ages of Poetry. In P. B. Shelley and T. L. Peacock, *A Defence of Poetry/The Four Ages of Poetry* (pp. 3–21). J. E. Jordan, Ed. Indianapolis: Bobbs-Merrill.

Penfield, W., and Rasmussen, T. (1950). *The Cerebral Cortex of Man: A Clinical Study of Localization of Function*. New York: Macmillan.

Peretz, I., Gagnon, L., Hébert, S., and Macoir, J. (2004). Singing in the Brain: Insights from Cognitive Neuropsychology. *Music Perception*, 21: 373–390.

Peretz, I., and Hyde, K. L. (2003). What is Specific to Music Processing? Insights from Congenital Amusia. *Trends in Cognitive Sciences*, 7: 363–367.

Peuser, G., and Temp, K. (1981). The Evolution of Jargonaphasia. In J. W. Brown (Ed.), *Jargonaphasia: Perspectives in Neurolinguistics, Neuropsychology, and Psycholinguistics* (pp. 259–293). New York: Academic Press.

Pinker, S. (1994). *The Language Instinct*. New York: William Morrow.

Pollak, T. A., Mulvenna, C. M., and Lythgoe, M. F. (2007). De novo Artistic Behaviour Following Brain Injury. In J. Bogousslavsky and M. G. Hennerici (Eds.), *Neurological Disorders in Famous Artists* (pp. 75–88). Basel: Karger.

Port, R. F., and Van Gelder, T. (1995). *Mind as Motion: Explorations in the Dynamics of Cognition*. Cambridge, MA: MIT Press.

Prokop, H. (1979). Eine schöpferische Produktion des Unbewussten. *Musik und Medizin*, 11: 49–56.

Prolo, P., and Licinio, J. (2002). DRD4 and Novelty Seeking. In J. Benjamin, R. P. Ebstein, and R. H. Belmaker (Eds.), *Molecular Genetics and the Human Personality* (pp. 91–108). Washington, D.C.: American Psychiatry Publishing.

Proust, M. ([1913–1927] 1966–1969). *À La recherche du temps perdu*. 3 Vols. Paris: Pléiade.

Puccetti, R. (1973). *The Trial of John and Henry Norton*. London: Hutchinson.

Pulli, K., Karma, K., Norio, R., Sistonen, P., Göring, H. H. H., and Järvelä, I. (2008). Genome-Wide Linkage Scan for Loci of Musical Aptitude in

Finnish Families: Evidence for a Major Locus at 4q22. *Journal of Medical Genetics*, 45: 451–456.

Purcell, H. ([1689] n.d.). *Dido and Aeneas*. E. J. Dent, Ed. London: Oxford University Press.

Radulescu, H. (2003). Brain and Sound Resonance: The World of Self-Generative Functions as a Basis of the Spectral Language of Music. In G. Avanzini, C. Faienza, D. Minciacchi, L. Lopez, and M. Majno (Eds.), *The Neurosciences and Music* (pp. 322–363). New York: New York Academy of Sciences.

Ramachandran, V. S., Azoulai, S., Stone, L., Srinivasan, A. V., and Bijoy, N. (2005). Grasping Metaphors and Thinking with Pictures: How Brain Damage Might Affect Thought and Language. Poster Presented at the 12th Annual Meeting of the Cognitive Neuroscience Society, New York, NY.

Ramachandran, V. S., and Hirstein, W. (1999). The Science of Art: A Neurological Theory of Aesthetic Experience. *Journal of Consciousness Studies* 6(6/7): 15–51.

Ramón y Cajal, S. (2001). *Vacation Stories: Five Science Fiction Tales*. Urbana: University of Illinois Press.

Rauch, A. (2000). *The Hieroglyph of Tradition: Freud, Benjamin, Gadamer, Novalis, Kant*. Cranbury, NJ: Associated University Presses.

Regard, M., and Landis, T. (1988). Beauty May Differ in Each Half of the Eye of the Beholder. In I. Rentschler, B. Herzberger, and D. Epstein (Eds.), *Beauty and the Brain: Biological Aspects of Aesthetics* (pp. 243–256). Basel: Birkhäuser.

Ribot, T. (1885). *La Psychologie allemande contemporaine (école expérimentale)*. Paris: Félix Alcan.

Richards, I. A. (1965). *The Philosophy of Rhetoric*. New York: Oxford University Press.

Richardson, A. (2001). *British Romanticism and the Science of the Mind*. Cambridge, UK: Cambridge University Press.

Rimbaud, A. (1967). "Voyelles." In *Oeuvres complètes* (p. 103). Paris: Pléiade.

Robbe-Grillet, A. (1963). *Pour un nouveau roman*. Paris: Gallimard NRF.

Robinson, D. (2007). Conclusion. In M. Bennett, D. Dennett, P. Hacker, J. Searle, and D. Robinson, *Neuroscience and Philosophy: Brain, Mind, and Language* (pp. 171–193). New York: Columbia University Press.

Rockwell, W. T. (2005). *Neither Brain nor Ghost: A Nondualist Alternative to the Mind-Brain Identity Theory*. Cambridge, MA: MIT Press.

Roethke, T. ([1951] 1966). "The Waking." In *The Collected Poems of Theodore Roethke* (p. 108). New York: Doubleday.

Rose, F. C. (Ed.). (2004). *Neurology of the Arts: Painting, Music, Literature*. London: Imperial College.

Rosenfeld, I., and Ziff, E. (2008). How the Mind Works: Revelations. *New York Review of Books*, June 26, 63–65.

Ross, D., Choi, J., and Purves, D. (2007). Musical Intervals in Speech.

Proceedings of the National Academy of Sciences of the United States of America, 104 (23): 9852–9857.

Rotenberg, V. S. (2004). The Peculiarity of the Right-Hemisphere Function in Depression: Solving the Paradoxes. *Progress in Neuropsychopharmacology and Biological Psychiatry*, 1: 1–13.

Rothenberg, D. (2005). *Why Birds Sing: A Journey through the Mystery of Bird Song*. New York: Basic Books.

Rousseau, G. S. (2004). *Nervous Acts: Essays on Literature, Culture and Sensibility*. Houndmills, UK: Palgrave Macmillan.

Rousseau, J.-J. ([1781] 1966). Essay on the Origin of Languages which Treats of Melody and Musical Imitation. In J. H. Moran and A. Gode (Trans.), *On the Origin of Language: Jean-Jacques Rousseau, "Essay on the Origin of Languages;" Johann Gottfried Herder, "Essay on the Origin of Language"* (pp. 1–74). Chicago: University of Chicago Press.

Rowland, I. D. (2006). The Titan of Titans. *New York Review of Books*, April 27, 8–11.

Rubin, D. C., and Greenberg, D. L. (2003). The Role of Narrative in Recollection: A View from Cognitive Psychology and Neuropsychology. In G. D. Fireman, T. E. McVay, Jr., and O. J. Flanagan (Eds.), *Narrative and Consciousness: Literature, Psychology, and the Brain* (pp. 53–85). New York: Oxford University Press.

Sacks, O. (1987). *The Man Who Mistook His Wife for a Hat and Other Clinical Tales*. New York: Harper and Row.

———. (2007). *Musicophilia: Tales of Music and the Brain*. New York: Alfred A. Knopf.

Salzman, M. (2000). *Lying Awake*. New York: Alfred A. Knopf.

Sapolsky, R., and Ehrlich, P. (2003). Appeal of the Rare. *Discover*, November, 60–63.

Saxe, R. (2005). Against Simulation: The Argument from Error. *Trends in Cognitive Science* 9(4): 174–179.

Scarry, E. (2004). Imagining Flowers: Perceptual Mimicry (Particularly Delphinium). In C. Armstrong and C. de Zegher (Eds.), *Ocean Flowers: Impressions from Nature* (pp. 243–268). Princeton: Princeton University Press.

Scherner, K. A. (1861). *Das Leben des Traums*. Berlin: Schindler.

Schleifer, R. (2001). The Poetics of Tourette Syndrome: Language, Neurobiology, and Poetry. *New Literary History*, 32: 563–584.

Schoenberg, A. (1975). *Style and Idea: Selected Writings of Arnold Schoenberg*. New York: St. Martins.

Schön, D., Gordon, R. L., and Besson, M. (2005). Musical and Linguistic Processing in Song Perception. *Annals of the New York Academy of Sciences*, 1060: 71–81.

Schubart, H. F. D. (1968). *Gedichte*. Frankfurt: Fischer Bücherei.

Seeley, W. P. (2005). Naturalizing Aesthetics: Art and the Cognitive Neuroscience of Vision. *Journal of Visual Arts Practice*, 5(3): 195–213.

Seeley, W. W., Matthews, B. R., Crawford, R. K., Gorno-Tempini, M. L., Foti, D., Mackenzie, I. R., and Miller, B. L. (2008). Unravelling Boléro: Progressive Aphasia, Transmodal Creativity and the Right Posterior Cortex. *Brain*, 131: 39–49.

Shaffir, M. (1940). *A Stezhke*. Montreal: Privately Printed.

Shakespeare, W. ([1608] 1948). *The Tragedy of King Lear*. In *Shakespeare: 23 Plays and the Sonnets* (pp. 784–827). New York: Harcourt Brace.

———. ([1609] 1948). *The Tragedy of Troilus and Cressida*. In *Shakespeare: 23 Plays and the Sonnets* (pp. 659–699). New York: Harcourt Brace.

———. ([1622] 1983). *Othello*. Harmondsworth: Penguin.

———. ([1623] 1948). *The Tragedy of Antony and Cleopatra*. In *Shakespeare: 23 Plays and the Sonnets* (pp. 867–908). New York: Harcourt Brace.

Shammi, P., and Stuss, D. T. (1999). Humour Appreciation: A Role of the Right Frontal Lobe. *Brain*, 122: 657–666.

Shelley, P. B. ([1817] 1956). "Hymn to Intellectual Beauty." In *Percy Bysshe Shelley: Collected Poems* (pp. 1–4). New York: Appleton-Century-Crofts.

———. ([1821] 1965). A Defence of Poetry. In P. B. Shelley and T. L. Peacock, *A Defence of Poetry/The Four Ages of Poetry* (pp. 23–80). J. E. Jordan, Ed. Indianapolis: Bobbs-Merrill.

Shepard, R. N. (1978). Externalization of Mental Images and the Act of Creation. In B. S. Randhawa and W. E. Coffman (Eds.), *Visual Learning, Thinking, and Communication* (pp. 133–189). New York: Academic Press.

Shlain, L. (1991). *Art and Physics: Parallel Visions in Space, Time, and Light*. New York: William Morrow.

Skoyles, J. R. (1998). Motor Perception and Anatomical Realism in Classical Greek Art. *Medical Hypotheses*, 51: 69–70.

Smith, B. H. (1988). *Contingencies of Value*. Cambridge, MA: Harvard University Press.

———. (2005). *Scandalous Knowledge: Science, Truth and the Human*. Edinburgh: Edinburgh University Press.

Smith, J. D., Shields, W. E., Schull, J., and Washburn, D. (1997). The Uncertain Response in Humans and Animals. *Cognition*, 62(1): 75–97.

Snow, C. P. (1959). *The Two Cultures and the Scientific Revolution*. New York: Cambridge University Press.

Snyder, A., and Mitchell, D. J. (1999). Is Integer Arithmetic Fundamental to Mental Processing?: The Mind's Secret Arithmetic. *Proceedings of the Royal Society of London, Series B: Biological Sciences*, 266: 587–592.

Solms, M. (2000). Dreaming and REM Sleep are Controlled by Different Mechanisms. *Behavioral and Brain Sciences*, 233: 843–850.

Solms, M., and Turnbull, O. (2002). *The Brain and the Inner World: An Introduction to the Neuroscience of Subjective Experience*. New York: Other Press.

Solso, R. L. (2003). *The Psychology of Art and the Evolution of the Conscious Brain*. Cambridge, MA: MIT Press.

Springer, U. S., Bowers, D., Goodman, W. K., Shapira, N. A., Foote, K. D., and Okun, M. S. (2006). Long-Term Habituation of the Smile Response with Deep Brain Stimulation. *Neurocase*, 12: 191–196.

Stafford, B. M. (2007). *Echo Objects: The Cognitive Work of Images*. Chicago: University of Chicago Press.

Stapp, H. (2007). Quantum Mechanical Theories of Consciousness. In M. Velmans and S. Schneider (Eds.), *The Blackwell Companion to Consciousness* (pp. 300–312). Malden, MA: Blackwell.

States, B. (1988). *The Rhetoric of Dreams*. Ithaca: Cornell University Press.

Stebbing, P. (2004). A Universal Grammar for Visual Composition? *Leonardo*, 37(1): 63–70.

Stein, B. E., Wallace, M. T., and Stanford, T. R. (2000). Merging Sensory Signals in the Brain: The Development of Multisensory Integration in the Superior Colliculus. In M. Gazzaniga (Ed.), *The New Cognitive Neurosciences* (pp. 55–71). Cambridge, MA: MIT Press.

Steinke, W. R., Cuddy, L. L., and Jakobson, L. S. (2001). Dissociations among Functional Subsystems Governing Melody Recognition after Right-Hemisphere Damage. *Cognitive Neuropsychology*, 18: 411–437.

Stella, F. (1986). *Working Space*. Cambridge, MA: Harvard University Press.

Stenzl, J. (1991 October). *Traum und Musik*. In H. K. Metzger and R. Riehn (Eds.), *Musik und Traum* (8–102). Musik-Konzepte, Vol. 74. Munich: Text+Kritik.

Stewart, L., Walsh, V., Frith, U., and Rothwell, J. (2001). Transcranial Magnetic Stimulation Produces Speech Arrest but Not Song Arrest. In R. J. Zatorre and I. Peretz (Eds.), *The Biological Foundations of Music*, (pp. 433–435). New York: New York Academy of Sciences.

Storr, A. (1992). *Music and the Mind*. New York: Free Press.

Streich, H. (1980). Musik im Traum. *Musiktherapeutische Umschau*, 1: 9–19.

Stuckey, C. F. (1993). Monet Revised? *Art in America*, November, 41–47.

Svoboda, E. (2006). Laser Lights Up Fruit-Fly Brains. *Discover*, January, 38.

Swirski, P. (2007). *Of Literature and Knowledge: Explorations in Narrative Thought Experiments, Evolution, and Game Theory*. New York: Routledge.

Taine, H. (1877). Les Vibrations cérébrales et la pensée. *Revue philosophique de la France et de l'étranger*, 2(1): 1–9.

Tallis, R. (2008). License my roving hands. *Times Literary Supplement*, April 11, 13–15.

Tambling, J. (2000). Operatic Sound and the Aural Unconscious. *Ars Lyrica*, 11: 23–38.

Tancredi, L. R. (2005). *Hardwired Behavior: What Neuroscience Reveals about Morality*. Cambridge, UK: Cambridge University Press.

Tankersley, D., Stowe, C. J., and Huettel, S. (2007). Altruism is Associated with an Increased Neural Response to Agency. *Nature Neuroscience*, 10(2): 150–151.

Taylor, J. (2007). *Science and Omniscience in Nineteenth-Century Literature*. Brighton: Sussex Academic Press.

Thaut, M. H. (2005). *Rhythm, Music, and the Brain: Scientific Foundations* 209
and Clinical Applications. New York: Routledge.

"timid bull fears mouse electrical stimulation of amygdala." This Web entry
brings up "Unit 1," under which, in "Unit 3," one reads, "a mouse receiving electrical stimulation to the amygdala will attack a cat. And sup-
pressing the amygdala will stop a bull in his tracks." http://itech.pjc.edu/
pgram/general/TestReview1.htm (accessed March 10, 2007).

Tolstoy, L. ([1862] 1965). *The Cossacks.* In *Leo Tolstoy: Short Novels* (pp. 279–
455). New York: Modern Library.

———. ([1865-1869] 1991). *War and Peace.* New York: Oxford University Press.

Tramo, M. J. (2001). Music of the Hemispheres. *Science,* 291: 54–56.

Tramo, M. J., Cariani, P. A., Koh, K. C., and Braida, L. D. (2005). Neurophysiology and Neuroanatomy of Pitch Perception: Auditory Cortex.
In G. Avanzini, S. Koelsch, L. Lopez, and M. Majno (Eds.). *The Neurosciences and Music II: From Perception to Performance* (pp. 148–174).
New York: New York Academy of Sciences.

Turner, F., and Pöppel, E. (1988). Metered Poetry, the Brain, and Time. In
I. Rentschler, B. Herzberger, and D. Epstein (Eds.), *Beauty and the Brain:
Biological Aspects of Aesthetics* (pp. 71–90). Basel: Birkhäuser.

Uga, V., Lemut, M. C., Zampi, C., Zilli, I., and Salzarulo, P. (2006). Music in
Dreams. *Consciousness and Cognition,* 15: 351–357.

Unger, D. N. S. (2005). Retinal Ennui. *Harvard Magazine,* November-
December, 11–12.

Uttal, W. R. (2001). *The New Phrenology: The Limits of Localizing Cognitive
Processes in the Brain.* Cambridge, MA: MIT Press.

Valéry, P. ([1920] 1957). "Le Cimetière marin." In *Oeuvres,* Vol. 1 (pp. 147–
151). Paris: Pléiade.

Vartanian, O., and Goel, V. (2004). Neuroanatomical Correlates of Aesthetic Preference for Paintings. Neuroreport 15(5): 893–897.

Vigny, A. de ([1832] 1963). *Stello: A Session with Doctor Noir.* Montreal:
McGill University Press.

———. ([1843] 1948). "La Flûte." In *Oeuvres complètes,* Vol. 1 (pp. 199–202).
Paris: Pléiade.

Vitruvius Pollio (1931–1934). *Vitruvius on Architecture.* 2 vols. Loeb Classical
Library. London: W. Heinemann.

Volkelt, J. (1875). *Die Traum-Phantasie.* Stuttgart: Meyer and Zeller.

Vygotsky, L. S. (1970). *Thought and Language.* Cambridge, MA: MIT Press.

Wade, N. (2005). Exploring a Hormone for Caring. *New York Times,* November 22, D5.

———. (2006). In Memory-Bank "Dialogue," the Brain is Talking to Itself.
New York Times, December 18, A21.

Wagner, C. (1978–1980). *Cosima Wagner's Diaries.* 2 Vols. New York: Harcourt Brace Jovanovich.

Wallin, N. L. (1991). *Biomusicology: Neurophysiological, Neuropsychologi-*

cal, and Evolutionary Perspectives on the Origins and Purposes of Music.
Stuyvesant, NY: Pendragon.

Warren, J. (2004). The Amusias. In F. C. Rose (Ed.), *Neurology of the Arts: Painting, Music, Literature* (pp. 275–305). London: Imperial College Press.

Weinberger, N. M. (2004). Music and the Brain. *Scientific American,* November, 88–96.

Weinstein, E. A. (1981). Behavioral Aspects of Jargonaphasia. In J. W. Brown (Ed.), *Jargonaphasia: Perspectives in Neurolinguistics, Neuropsychology, and Psycholinguistics* (pp. 139–149). New York: Academic Press.

Whitelaw, M. (2004). *Metacreation: Art and Artificial Life.* Cambridge, MA: MIT Press.

Williams, L. (1980). *The Dancing Chimpanzee.* London: Allison and Busby.

Wilson, E. A. (2004). *Psychosomatic.* Durham, NC: Duke University Press.

Wilson, R. A. (2004). *Boundaries of the Mind.* Cambridge, UK: Cambridge University Press.

Wittgenstein, L. (1968). *Philosophical Investigations.* New York: Macmillan.

Wordsworth, W. ([1800] 1950). Preface to the "Lyrical Ballads." In *Selected Poems* (pp. 1–2). G. W. Meyer, Ed. New York: Appleton-Century-Crofts.

———. ([1807] 1965). "The Solitary Reaper." In *Selected Poems and Prefaces* (pp. 367–368). J. Stillinger, Ed. Boston: Houghton Mifflin.

Wróbel, J. (1990). *Language and Schizophrenia.* Amsterdam: John Benjamins.

Zaidel, D. W. (2005). *Neuropsychology of Art: Neurological, Cognitive, and Evolutionary Perspectives.* New York: Psychology Press.

Zaidel, E. (2001). Hemispheric Specialization for Language in the Split Brain. In R. S. Berndt (Ed.), *Language and Aphasia* (pp. 393–418). Handbook of Neuropsychology, 2nd ed., Vol. 3. Amsterdam: Elsevier.

Zeki, S. (1999). *Inner Vision: An Exploration of Art and the Brain.* New York: Oxford University Press.

Zilkens, U. (1997). *Franz Schubert: Vom Klavierlied zum Klavierquintett— Die Forelle im Spiegel ihrer Interpretationen durch Musiktheoretiker und Musiker.* Cologne: P. J. Tonger.

Zimmer, C. (2005). Neuron Network Goes Awry, and Brain Becomes an iPod. *New York Times,* July 12, F1 and F6.

———. (2006). Silent Struggle: A New Theory of Pregnancy. *New York Times,* March 14, D1 and D6.

Zunshine, L. (2006). *Why We Read Fiction.* Columbus: Ohio State University Press.

WORKS CONSULTED

Bentivoglio, M. (2003). Musical Skills and Neural Functions: The Legacy of the Brains of Musicians. In G. Avanzini, C. Faienza, D. Minciacchi, L.

Lopez, and M. Majno (Eds.), *The Neurosciences and Music* (pp. 234–243). 211
New York: New York Academy of Sciences.

Bernstein, C. (1994). Semblance. In Paul Hoover (Ed.), *Postmodern American Poetry* (pp. 676–679). New York: W. W. Norton.

Biology of Music (2000). *Economist*, February 12, 83–85.

Black, M. (1962). *Models and Metaphors*. Ithaca: Cornell University Press.

Bosma, H. (2002). Wonderful Music. *Dream Time*, 19: 6–9.

Brandt, L., and Brandt, P. A. (2005). Making Sense of a Blend: A Cognitive-Semiotic Approach to Metaphor. *Annual Review of Cognitive Linguistics*, 3: 216–249.

Brown, J. W. (1989). Essay on Perception. In J. W. Brown (Ed.), *Neuropsychology of Visual Perception* (pp. 233–255). Hillsdale, NJ: Lawrence Erlbaum.

Cipolotti, L., and Van Harskamp, N. (2001). Disturbances of Number Processing and Calculation. In R. S. Berndt (Ed.), *Language and Aphasia* (pp. 305–331). Handbook of Neuropsychology, 2nd ed., Vol. 3. Amsterdam: Elsevier.

Clarke, B., and Henderson, L. D. (Eds.) (2002). *From Energy to Information: Representation in Science and Technology, Art, and Literature*. Stanford: Stanford University Press.

Dierig, S. (2001). Con Sordino for Piano and Brain: Bohemian Neuroscience in a 1900 Cultural Metropolis. *Configurations*, 9: 413–440.

Fukui, H. (2001). Music and Testosterone. In R. J. Zatorre and I. Peretz (Eds.), *The Biological Foundations of Music* (pp. 448–451). New York: New York Academy of Sciences.

Glausiusz, J. (2001). The Genetic Mystery of Music. *Discover*, August, 70–75.

Goguen, J. A. (1999). Art and the Brain: Editorial Introduction. *Journal of Consciousness Studies*, 6: 5–14.

Gombrich, E. H. (1965). *Art and Illusion: A Study in the Psychology of Pictorial Representation*. New York: Bollingen Foundation.

Gottschall, J., and Wilson, D. S. (Eds.) (2005). *The Literary Animal: Evolution and the Nature of Narrative*. Evanston: Northwestern University Press.

Hodges, D. A., Hairston, W. D., and Burdette, J. H. (2006). Aspects of Multisensory Perception: The Integration of Visual and Auditory Information in Musical Experiences. In G. Avanzini, S. Koelsch, L. Lopez, and M. Majno (Eds.), *The Neurosciences and Music II: From Perception to Performance* (pp. 175–185). New York: New York Academy of Sciences.

Janata, P., Birk, J. L., Van Horn, J. D., Leman, M., Tillmann, B., and Bharucha, J. (2002). The Cortical Topography of Tonal Structures Underlying Western Music. *Science*, 298: 2167–2170.

Kayser, C. (2007). Listening with your Eyes. *Scientific American Mind*, April/May, 24–29.

Lake, N. (2002). Spinal Chords: Cortices in C Minor. *Harvard Magazine*, March-April, 16–18.

212   Leutwyler, K. (2001). Exploring the Musical Brain. *SciAm.com*, January 22.
      http://www.sciam.com/article.cfm?id=exploring-the-musical-bra
      (accessed December 17, 2008).

Malabou, C. (2008). *What Should We Do with Our Brain?* New York: Ford-
ham University Press.

Michael, C. R. (1969). Retinal Processing of Visual Images. *Scientific American* May, 105–114.

National Research Council (2008). *Emerging Cognitive Neuroscience and
Related Technologies.* Washington, D.C.: National Academies Press.

Oberman, L. M., and Ramachandran, V. S. (submitted). Evidence for Deficits in Mirror Neuron Function, Multisensory Integration, and Sound-
Form Symbolism in Autism Spectrum Disorders. Invited submission to
special issue of *Social Neuroscience* on Mirror Neurons and Communication.

Onians, J. (2007). *Neuroarthistory: From Aristotle and Pliny to Baxandall
and Zeki.* New Haven: Yale University Press.

Rizzolatti, G., and Sinigaglia, C. (2008). *Mirrors in the Brain: How Our
Minds Share Actions and Emotions.* Oxford: Oxford University Press.

Samson, S., and Zatorre, R. (1991). Recognition Memory for Text and Melody of Songs After Unilateral Temporal Lobe Lesion: Evidence for Dual
Encoding. *Journal of Experimental Psychology: Learning, Memory, Cognition*, 17: 793–804.

Wade, N. (2003). We Got Rhythm: The Mystery is How and Why. *New York
Times*, September 16, D1 and D4.

Zimmer, C. (2005). Looking for Personality in Animals, of All People. *New
York Times*, March 1, D1 and D6.

Index

www.ingramcontent.com/pod-product-compliance
Lightning Source LLC
Chambersburg PA
CBHW071417180526
45170CB00001B/139